Key Maths Skills

© C.R. Draper, 2022

The answers are available online at: http://www.warrupress.com/education

All rights reserved. No part of this book may be reproduced or transmitted in any form or by any means without written permission of the author.

Published: Warru Press, 2022

ISBN: 978-1-922819-00-0

Contents

Contents 2

Chapter 1: Basic Number 8
    1.1 Place Value 8
    1.2 Bigger and smaller 11
    1.3 Negative Numbers 12
    1.4 Rounding 13

Chapter 2: Addition and Subtraction 16
    2.1 Addition 16
    2.2 Adding Decimals and Missing Numbers 19
    2.3 Subtraction 21
    2.4 Subtracting Decimals and Missing Numbers 24
    2.5 Addition and Subtraction are Opposites 26

Chapter 3: Multiplication and Division 28
    3.1 Multiplication and division by tens 28
    3.2 Long Multiplication 30
    3.3 Short division 33
    3.4 Long Division 37

Chapter 4: Further calculations 45
    4.1 Addition and subtraction of negatives 45
    4.2 BIMDAS 48
    4.3 Inverse problems 49
    4.4 Magic Squares 51
    Puzzle Page 1 55

Chapter 5: More Number 56
    5.1 Factors 56
    5.2 Multiples 59
    5.3. Prime numbers 60
    5.4 Squares and square roots 61

Chapter 6: Number Relationships 63
    6.1 Number lines 63
    6.2 Indices 65
    6.3 Highest common factor (HCF) 66
    6.4 Lowest common multiple (LCM) 68

Chapter 7: Fractions 71
- 7.1 Simple fractions 71
- 7.2 Equivalent fractions 73
- 7.3 Simplifying fractions 75
- 7.4 Improper fractions and mixed numbers 76
- 7.5 Lowest Common Denominator (LCD) 80

Chapter 8: Calculating Fractions 82
- 8.1 Adding fractions 82
- 8.2 Subtracting Fractions 83
- 8.3 Multiplying Fractions 84
- 8.4 Dividing Fractions 86
- 8.5 Calculating with mixed numbers 87

Chapter 9: Solving Fraction Problems 91
- 9.1 Complementary fractions 91
- 9.2 Fraction of a number 92
- 9.3 Working out the whole or original amount 93
- 9.4 Fraction Problems 95

Chapter 10: Percentages 97
- 10.1 Percentages 97
- 10.2 Percentage to fraction 99
- 10.3 Fraction to percentage 100
- 10.4 Decimals and percentages 101
- 10.5 Percentage of an amount 103

Chapter 11 Calculating with Percentages 105
- 11.1 Calculating the amount 105
- 11.2 Increasing and decreasing by a percentage 107
- 11.3 Original Number 110
- 11.4 Percent increase and decrease 111

Chapter 12: Decimals 114
- 12.1 Decimal multiplication 114
- 12.2 Division of decimals 115
- 12.3 Decimals to fractions 117
- 12.4 Fractions to decimal 119
- 12.5 Decimals, Fractions and Percentages 120

Puzzle Page 2 122

Chapter 13: Looking at data 123

    13.1 Mean 123

    13.2 Mean problems 124

    13.3 Mode and Median 127

    13.4 Range 129

Chapter 14: Money 130

    14.1 Dollars and cents 130

    14.2 Calculating with money 131

    14.3 Costs 133

    14.4 Currency conversions 135

Chapter 15: Time 137

    15.1 As time goes by 137

    15.2 12 hour clock 138

    15.3 24 Hour Time 141

    15.4 Time fractions 142

Chapter 16: Calculating Time 143

    16.1 Adding time 143

    16.2 Subtracting Time 147

    16.3 Multiplying time 151

    16.4 Dividing Time 153

    16.5 Once upon a time 155

Chapter 17: Time Problems 157

    17.1 Centuries 157

    17.2 Logic time 158

    17.3 Timetables 161

    17.4 Time zones 162

    Puzzle Page 3 164

Chapter 18: Measurement 165

    18.1 Metric Units 165

    18.2 Metric Conversions 166

    18.3 Imperial Units 167

    18.4 Metric – Imperial Units 168

Chapter 19: Using Measurement 169

19.1 Using scales 169

19.2 Temperature 172

19.3 Calculating with measurements.. 174

19.4 Coordinates 176

Chapter 20: Lines

20.1 Types of line 178

20.2 Lines and Angles.. 181

20.3 Alternate, Corresponding and Opposite Angles 183

20.4 Triangles and Quadrilaterals 185

Chapter 21: 2D Shapes

21.1 Polygons 187

21.2 Triangles and Quadrilaterals 189

21.3 Circles and Ellipses 192

Chapter 22: Area and Perimeter

22.1 Perimeter 195

22.2 Area of a rectangle 198

22.3 Area of a triangle... 201

22.4 Complex Area 204

Chapter 23: 2D Shape Problems

23.1 Reverse Area and Perimeter Problems 209

23.2 Area and perimeter 212

23.3 Reflective Symmetry 214

23.4 Rotational Symmetry 216

Chapter 24: 3D Shapes

24.1 Prisms and Pyramids 218

24.2 Volume 222

24.3 Surface Area 225

24.4 Nets 227

Puzzle Page 4 228

Chapter 25: Navigation

25.1 Compass Points 229

25.2 Directions 231

25.3 Speed 232

25.4 Distance-time graphs 234

Chapter 26: Ratios 236
26.1 Ratios 236
26.2 Ratio Problems 238
26.3 Maps and Scales 241
26.4 Hidden Ratios 244

Chapter 27: Probability 247
27.1 What is Probability 247
27.2 Calculating Probability 248
27.3 Outcomes of Combined Events 251
27.4 Probability of Combined Events 255

Chapter 28: Tables and Graphs 258
28.1 Two Way Tables 258
28.2 Distance Tables 260
28.3 Bar Charts 262
28.4 Pie Charts 264

Chapter 29: Graphs and more 268
29.1 Line graphs 268
29.2 Depicting Data Pictorially 271
29.3 Venn Diagrams 274
29.4 Flow Charts 277

Chapter 30: Algebra 281
30.1 Representing Unknowns with Letters 281
30.2 Writing Math Problems using Algebra 283
30.3 Substitution 285
30.4 Using Formulae 286
30.5 Algebraic terms 290

Chapter 31: Using Algebra 293
31.1 Multiplying Terms 293
31.2 Dividing Terms 294
Combining multiplication and addition 295
31.3 Balancing an Equation 296
31.4 Creating Formulae 298

Puzzle Page 5 304

Chapter 1: Basic Number

1.1 Place Value

In our decimal system every place value to the left is worth ten times more than the place value to the right. This is true both sides of the decimal point.

millions	hundred thousands	ten thousands	thousands	hundreds	tens	units	.	tenths	hundredths	thousandths
1000 000	100 000	10 000	1000	100	10	1		1/10	1/100	1/1000

So, a seven in the thousands place is worth ten times more than a seven in the hundreds place.

To write a number, place the digits in the correct place value, putting zero in any unused place values.

Example 1: Write in digits the number four thousand and twenty three.

thousands	hundreds	tens	units
4	0	2	3

Answer: 4023

Example 2: Write in digits the number three hundredths.

units	.	tenths	hundredths
0	.	0	3

Answer: 0.03

Exercise 1.1a

Write the following numbers using digits.

1. Eleven _____
2. Fifty eight _____
3. Three hundred and one _____
4. Forty two thousand and six _____
5. Three hundred and sixty two thousand _____
6. One million three thousand and twenty two _____
7. Five million sixty thousand and forty five _____
8. Two tenths _____
9. Nine hundredths _____
10. Seven point two _____

Exercise 1.1b

Write the following numbers using words.

1. 27 _____
2. 13 _____
3. 432 _____
4. 3 502 _____
5. 26 023 _____
6. 103 004 _____
7. 1 654 000 _____
8. 2 002 002 _____
9. 0.1 _____
10. 0.007 _____

Decimals can be written as fractions, by placing the digits over the smallest place value.

> Examples: As 0.7 is seven in the tenths place, we can write it as the fraction $\frac{7}{10}$.
>
> In the number 0.63, the smallest digit is in the hundredths place, so it can be written as $\frac{63}{100}$.
>
> This works because every place value is worth ten times more, so the 6 is worth ten times a hundredth.

Exercise 1.1c

Complete the following.

1. $0.1 = \frac{}{10}$

2. $0.09 = \frac{9}{}$

3. $26.33 = 20 + \underline{} + \frac{33}{100}$

4. $3.27 = 3 + \frac{}{10} + \frac{7}{100}$

5. $5.68 = 5 + \frac{6}{10} + \frac{}{100}$

6. $4.102 = 4 + \frac{1}{10} + \frac{2}{}$

7. $96.56 = 90 + 6 + \frac{}{10} + \frac{}{100}$

8. $96.56 = 96 + \frac{56}{}$

9. $34.107 = 34 + \frac{107}{}$

10. $9.173 = 9 + \frac{}{1000}$

1.2 Bigger and smaller

To work out if a number is bigger or smaller than another number look at the largest place value first. The number which has the biggest digit in this place value will be the biggest number. If the digits are the same in this place value, move to the next biggest place value and so on.

> For example, which is bigger 2.34 or 2.304
>
> The biggest place value is the units.
> Both numbers have a two in this place value, so we look at the next place value, the tenths.
> Both numbers have a three in the tenths place so we move to the next place value, the hundredths.
> The first number has a four in the hundredths while the second number has a zero.
> Four is bigger than zero so 2.34 is bigger than 2.304.

The symbol > can be used to represent greater than and < can be used to represent less than. All you need to remember is that the big end points to the biggest number and the small end points to the smallest number. Or you may prefer to think of a greedy, hungry crocodile that always goes for the largest number. So we could write our answer to the above example as 2.34 > 2.304.

Exercise 1.2

Rearrange the numbers below in size order, largest first.

1. 3 222   521   5721   11   6032   6201

   _____   _____ _____ _____ _____ _____

2. 5781   9067   345   2001   999   1002

   _____   _____ _____ _____ _____ _____

3.   0.01   0.2   1.3   1.001   0.009   0.04

   ____   ____   ____   ____   ____   ____

Which is more?

4. 326   or   309   _____
5. 99   or   101   _____
6. 0.3   or   0.199   _____

Put a < or > between the numbers below.

7. 34 ____ 52
8. 803 ____ 789
9. 0.34 ____ 0.211
10. 0.02 ____ 0.001

1.3 Negative Numbers

A negative number is a number less than zero.

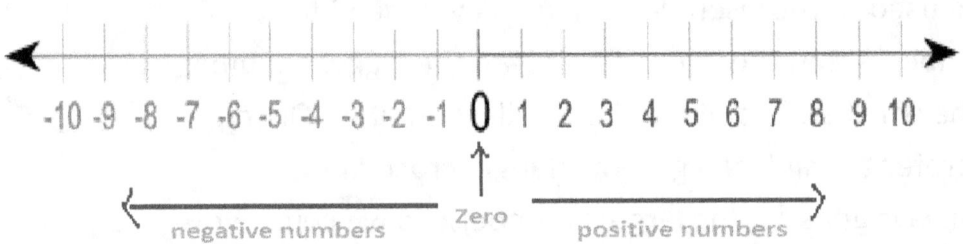

So -1 is one number less than zero.

Temperature is a common use of negative numbers.

Exercise 1.3

What number is?

1. Three less than zero? ____

2. 12 less than zero? ____

3. Two less than -4? ____

4. Four less than one? ____

5. Order these numbers starting from the smallest.
   2, -2, 1, -1, 0

   ____ ____ ____ ____ ____

Put < or > between the following temperatures.

6. -4°C ____ -2°C

7. -16°C ____ -17°C

8. -20°C ____ 20°C

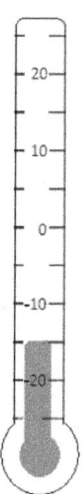

9. The temperature is -9°C. How much must it rise to reach 3°C? ____

10. What temperature does the thermometer show? ____

1.4 Rounding

To round
1. Look at the place value to the right
2. Ask "Is it 5 and above or below 5"
3. If it's 5 or above – round up
4. It it's below 5 – it stays as it is.

Example 1: Round 328 to the nearest 10.
- The next digit is 8, which is 5 or above, so we round up
- The answer is 330

Example 2: Round 328 to the nearest hundred
- The next digit is 2, which is less than 5, so the number stays the same
- The answer is 300.

Example 3: Round 795 to the nearest ten.
- The next digit is 5, which is 5 or above, so we round up
- When 9 is rounded up we get 10. So we write a zero and add one to the next place value.
- The answer is 800.

Rounding decimals works in the same way as rounding any other number, except that the place values that are rounded are no longer written.

Example 4: Round 4.827 to two decimal places
- The next digit is 7 which is 5 or above, so we round up
- So the answer is 4.83

Exercise 1.4

Round to the nearest 10

1. 57 _____
2. 35 _____
3. 472 _____
4. 295 _____

Round to the nearest whole number

5. 37.1 _____
6. 48.7 _____
7. 55.5 _____
8. 39.6 _____

Round to two decimal places

9. 38.724 _____

10. 49.381 _____

Exercise 1.5 Practice

1. In the number 857 392 what digit is in the ten thousands place? _____

2. In the number 163.582 what digit is in the hundredths place? _____

3. Write five hundred and two thousand and sixty two in digits. _____

4. Underline the number which is the largest.
   382   392   389   390.2

5. Underline the number which is the smallest.
   0.01   0.001   0.101   0.0009

6. Write the number thirty below zero. _____

7. Take 4 from -3 _____

8. Round 72 to the nearest 10 _____

9. Round 185 to the nearest 10 _____

10. Round 3.875 to 2 decimal places. _____

Chapter 2: Addition and Subtraction

2.1 Addition

Words for addition include: add, sum, total, increase, plus and combine.

When adding larger numbers write the numbers in columns, according to their place value.

$$\begin{array}{r} 3\ 8\ 1 \\ +\ 2\ 1\ 6 \\ \hline \end{array}$$

Then add each column in turn, starting with the smallest place value.

$$\begin{array}{r} 3\ 8\ 1 \\ +\ 2\ 1\ 6 \\ \hline 7 \end{array}$$ ← Start here

Continue with the next smallest place value.

$$\begin{array}{r} 3\ 8\ 1 \\ +\ 2\ 1\ 6 \\ \hline 5\ 9\ 7 \end{array}$$

If when we add a column the answer is more than 10, we write the units under that column and carry the tens to the next column.

$$\begin{array}{r} ^{1}\ \\ 5\ 8\ 7 \\ +\ 6\ 7\ 9 \\ \hline 6 \end{array}$$

7+9=16

So, I put down the 6 and carry the 1

Then continue as before.

$$\begin{array}{r} {\scriptstyle 11}\\ 5\ 8\ 7\\ +\ 6\ 7\ 9\\ \hline 1\ 2\ 6\ 6 \end{array}$$

Remember to write in your carrying.

Exercise 2.1

1.
$$\begin{array}{r} 2\ 4\ 7\\ +\ 6\ 5\ 2\\ \hline \end{array}$$

2.
$$\begin{array}{r} 5\ 8\ 9\\ +\ 4\ 5\ 6\\ \hline \end{array}$$

3.
$$\begin{array}{r} 9\ 2\ 8\\ +\ 3\ 8\ 6\\ \hline \end{array}$$

4.
$$\begin{array}{r} 1\ 2\ 3\ 6\\ 3\ 9\ 4\\ +\ 2\ 8\\ \hline \end{array}$$

5. What is the sum of 1356 and 644? _____

6. If Jack has 36 sticker cards, Joe has 92 and Jarred has 59. How many do they have altogether? _____

7. Sam and Simeon are packing glasses. There are 36 in one box, 45 in another and still 12 to pack. How many glasses are there altogether? _____

8. A shop sells 327 shirts in a week. There were 243 shirts left in the shop. There were no deliveries during the week. How many shirts were there at the beginning of the week? _____

9. Jo runs 14km on Monday, 12km on Wednesday and 19km on Friday. She does not run on Tuesday or Thursday. How many kilometres in total does she run?

_____

10. Seonaid and Laifa plant 362 Yew trees, 248 Sycamore trees and 520 Larch trees. How many trees do they plant in total? _____

2.2 Adding Decimals and Missing Numbers

When adding decimals, the decimal points go directly under each other.

```
    4 . 7 2 1
    3 . 6
+   9 . 4 2
```

If we have a different number of decimal places we may add zeros, to make them the same.

```
    4 . 7 2 1
    3 . 6 0 0
+   9 . 4 2 0
```

Then it works the same as before, starting from the smallest place value.

```
  1
    4 . 7 2 1
    3 . 6 0 0
+   9 . 4 2 0
   17. 7 4 1
```

If numbers are missing, work it out in the normal way from the smallest place value.

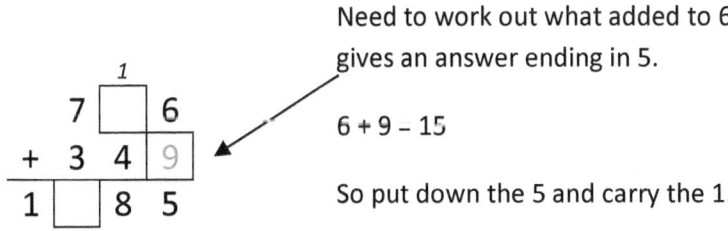

Need to work out what added to 6 gives an answer ending in 5.

6 + 9 – 15

So put down the 5 and carry the 1.

Make sure you remember to carry in the normal way.

Exercise 2.2

1.   1 . 3 2
      + 0 . 5 4

2.   3 . 2 7
      + 0 . 7 4

3.   7 . 2 7
      + 0 . 9

4. 3.64 + 5.8 = _____

5. 7.689 + 4.331 = _____

6. At the shops Ameera buys chicken costing $4.20 and $5.17 of lamb. How much did her meat cost? _____

7. Mrs Ifyano buys vegetables costing $12.36 and dairy products costing $6.75. How much does she spend? _____

8. John has three lengths of wood, measuring 67.3cm, 72.9cm and 6.7cm. What is the longest piece of wood he could make by gluing the three pieces together? (assume the glue adds no length). _____

9. Find the missing numbers.

        3 ☐ 4
     + 9 4 3
      1 ☐ 8 ☐

10. Find the missing numbers.

        8 ☐ 7
     + 8 9 ☐
      1 ☐ 8 4

2.3 Subtraction

Words for subtraction include: difference, subtract, take-away, take, reduce, minus, less or deduct.

As in addition we must write the numbers in columns so that the numbers with the same place value are underneath each other.

$$\begin{array}{r} 5\ 8\ 7 \\ -\ 1\ 3\ 4 \\ \hline \end{array}$$

Then we subtract each column in turn, starting from the smallest place value.

$$\begin{array}{r} 5\ 8\ 7 \\ -\ 1\ 3\ 4 \\ \hline 4\ 5\ 3 \end{array}$$

If the digit at the bottom is smaller than the digit at the top, we need to borrow.

$$\begin{array}{r} 6\ 2\ 4 \\ -\ 3\ 7\ 5 \\ \hline \end{array}$$

I cannot do 4-5 so I borrow 1 from the 2. The next place value is always worth ten times more, so the 1 we borrow is worth ten, so the 4 becomes 14.

$$\begin{array}{r} \ ^{1} \\ 6\ \cancel{2}\ ^{1}4 \\ -\ 3\ 7\ 5 \\ \hline 9 \end{array}$$

Can't do 4-5, so I borrow 1 from the 2. The 4 then becomes 14.
14-5=9

I can't do 1-7, so I borrow from the 6 in the same way.

$$\begin{array}{r} {}^{5}\cancel{6}{}^{11}\cancel{7}{}^{1}4 \\ -375 \\ \hline 249 \end{array}$$

If there are zeros at the top we need to borrow from the first non-zero number.

$$\begin{array}{r} {}^{3}\cancel{4}001 \\ -364 \\ \hline \end{array}$$

I need to borrow from the thousands which is worth 10 in the hundreds, but I immediately borrow from there leaving 9. I now have 10 in the tens place, but again I need to immediately borrow from there leaving 9 in the tens place and adding 10 to my units to give 11.

$$\begin{array}{r} {}^{3}\cancel{4}{}^{9}\cancel{0}{}^{9}\cancel{0}{}^{1}1 \\ -364 \\ \hline 3637 \end{array}$$

Exercise 2.3

1.
$$\begin{array}{r} 9\;2\;7 \\ -\;3\;1\;5 \\ \hline \end{array}$$

2.
$$\begin{array}{r} 1\;5\;7\;8 \\ -\;\;\;4\;2\;6 \\ \hline \end{array}$$

3.
$$\begin{array}{r} 1\;3\;8\;2 \\ -\;\;\;4\;0\;3 \\ \hline \end{array}$$

4.
$$\begin{array}{r} 2\;9\;7\;6 \\ -1\;9\;8\;4 \\ \hline \end{array}$$

5.
```
   3 0 0 0
 -   3 2 7
 ─────────
```

6. Find the difference between 1000 and 64. _____

7. Ricardo buys a shirt costing $57 but has a $15 off voucher. How much does the shirt cost him?
_____

8. A shop sells 527 cans of soft drink on a Saturday. If they started with 1000 cans, how many do they have left? _____

9. A hotel has 320 rooms. If 76 are booked how many are still available? _____

10. Reena has 126 stickers. If she gives 35 away to friends how many does she have left? _____

2.4 Subtracting Decimals and Missing Numbers

When subtracting decimals, zeros must be added so that each number has the same number of digits after the decimal point.

$$\begin{array}{r} 364.2 \\ -15.361 \\ \hline \end{array}$$

becomes

$$\begin{array}{r} 364.200 \\ -15.361 \\ \hline \end{array}$$

0-1 I can't do, so I borrow from the 2, my first non-zero number.

$$\begin{array}{r} \overset{5}{3}\overset{13}{6}\overset{}{4}.\overset{11}{2}\overset{9}{0}\overset{}{{}^{1}0} \\ -15.361 \\ \hline 348.859 \end{array}$$

If there are numbers missing work it out in the normal way starting with the smallest place value. Remember to borrow as normal.

$$\begin{array}{r} 9\ 8\ \square \\ -\ \square\ 8\ 5 \\ \hline 1\ \square\ 9 \end{array}$$

becomes

$$\begin{array}{r} \overset{8}{9}\ \overset{17}{8}\ {}^{1}4 \\ -\ 7\ 8\ 5 \\ \hline 1\ 9\ 9 \end{array}$$

Exercise 2.4

1.   1 . 7 5
     - 0 . 5 4
     ―――――

2.   3 . 4 2
     - 1 . 6 7
     ―――――

3.   1 . 4
     - 0 . 5 2
     ―――――

4.   1
     - 0 . 3 7
     ―――――

5. A piece of plastic has a length of 12.5cm. After David cuts some off it measures 7.3cm. How much did David cut off? _____

6. Alysha measures 3.1kg of flour and 2.7kg of sugar. How much more flour does she have than sugar? _____

7. Bertie bought 8.7 kg of flour and gave 7.4 kg of it to a neighbour. How much flour does Bertie have left? _____

8. Brett hiked 10.6 miles on Saturday and 8.7 miles on Sunday. How much further did Brett walk on Saturday than on Sunday? _____

9.   ☐ ☐ 7
    - 8 3 ☐
    ―――――
         8 4

10.   ☐ 4 2
     - 8 3 ☐
     ―――――
       1 ☐ 4

2.5 Addition and Subtraction are Opposites

If two numbers are added, the answer minus either number will be the other number.

So, if   384 + 512 = 896

then   896 − 512 = 384

and   896 − 384 = 512

This can be used to solve inverse problems, where you are given an answer and have to work back to the starting point.

> Example: 36 was added to a number and the answer was 60. What was the original number?
>
> 36 was added so to go the "other way" I will subtract 36.
>
> $$60 - 36 = 24$$
>
> So, my answer is 24.

Exercise 2.5

1. If 564 + 396 = 960, what is 960 − 564? _____

2. If 1054 − 389 = 665, what is 665 + 389? _____

3. If 472 + 529 = 1001, what is 1001 − 529? _____

4. If 37.86 − 29.78 = 8.08, what is 37.86 − 8.08? _____

5. Laura thinks of a number and adds 25, to get 57. What was the number? _____

6. Zoe started with a piece of wood. After she cut off 5.6cm she had 47cm remaining. What length did she start with? _____

7. Sanjana read 23 pages of her book. She had previously read 96 pages. If she still had 180 pages to read, how many pages does the book have? _____

8. Isaac bought some fish and chips. Altogether he paid $10.20. If the fish cost $8.75, how much did the chips cost? _____

9. When 27 is added to a number the answer is 56.
   What is the number? _____

10. If 45 stickers are given away leaving 96 stickers. How many stickers were there to begin with? _____

Chapter 3: Multiplication and Division

3.1 Multiplication and division by tens

To multiply whole numbers by 10, 100, 1000 etc, simply add the number of zeros in the multiplier onto the number we are multiplying.

Therefore, to multiply by:

    10   add 0

    100   add 00

    1000   add 000 etc

Since division is the opposite of multiplication to divide we simply subtract the same number of zeros.

To multiply decimals we move the decimal point to the right by the number of zeros.

Therefore, to multiply by:

    10   move the decimal point   1 place

    100   move the decimal point   2 places

    1000   move the decimal point   3 places etc.

To divide we move the decimal point the same number of places in the opposite direction.

Just remember with numbers greater than one, if you are multiplying the answer will be bigger and dividing it will be smaller.

Example: 3.567 x 100

We are multiplying by 100, so move the decimal 2 places, so that the number will be bigger.

$$3.567 \times 100 = 356.7$$

Example: 48.2 ÷ 100

We again move the decimal place two places, but because we are dividing we will move the decimal place to the left (so that the number is smaller).

$$48.2 \div 100 = 0.482$$

If when the decimal point is moved there are not enough digits, then add zeros as necessary.

Example: 3.6 x 1000

$$3.6 \times 1000 = 3.600 \times 1000 = 3600$$

Example: 5.7 ÷ 1000

$$5.7 \div 1000 = 005.7 \div 1000 = 0.0057$$

Exercise 3.1

1. 62 x 100 = _____
2. 342 000 ÷ 100 = _____
3. 48 x 10 = _____
4. 9 600 000 ÷ 10 000 = _____
5. 3.2478 x 100 = _____
6. 567.1 ÷ 10 = _____
7. 462 ÷ 1000 = _____
8. 3.23 x 1000 = _____
9. 4.5 ÷ 1000 = _____
10. 3.8 x 100 = _____

3.2 Long Multiplication

Words for multiplication include: product, multiply and times.

We start by multiplying the top number by the units of the bottom number, starting with the units of the top number.

Example: 132 x 23

```
    1 3 2
  x   2 3
  -------
        6
```

Start by multiplying 3 x 2.

3 x 2 = 6

Write the 6 underneath and move on to the next top digit.

3 x 3 = 9 and so on

```
    1 3 2
  x   2 3
  -------
    3 9 6
```

Then we move onto the tens place of the bottom number. Because the 2 is in the tens place we are actually multiplying by 20, so we add a zero and then multiply as before.

```
    1 3 2
  ×   2 3
  ─────────
    3 9 6        You must add this zero before
          0      multiplying by 2.
```

```
    1 3 2
  ×   2 3
  ─────────
    3 9 6        Then multiply as before
  2 6 4 0
  ─────────      Then add
  3 0 3 6
```

Once we have multiplied by all the digits in the bottom number, we add to get our answer. In this case our answer is 3036.

If, when I multiply two digits together the answer is greater than ten, I write down the units and carry the tens.

```
    5 8 7        7 × 4 = 28
  ×     4
  ─────────      So, I put down the 8 and carry the 2.
        ²8
```

then

```
    5 8 7        Then 8 × 4 = 32  PLUS
  ×     4        the 2 I carried equals 34.
  ─────────      So, I put down the 4 and carry the 3.
      ³4 ²8
```

then

```
    5 8 7        Then 5 × 4 = 20  PLUS
  ×     4        the 3 I carried = 23.
  ─────────      There are no more digits,
  2 3 ³4 ²8      so I write the 23.
```

Exercise 3.2

1. $\begin{array}{r} 57 \\ \times3 \\ \hline \end{array}$

2. $\begin{array}{r} 96 \\ \times7 \\ \hline \end{array}$

3. $\begin{array}{r} 329 \\ \times52 \\ \hline \end{array}$

    _____

4. Scarlett has four packs of 52 playing cards. How many cards does she have? _____

5. A sports' tournament has 17 teams of nine players competing. How many competitors are there? _____

6. If a school has 128 students in each year group and seven year groups in the school, how many students are there in the school? _____

7. In a cinema there are 32 rows of seats. If the seats in each row are labelled A to Z, how many seats are there altogether? _____

8. There are 17 sweets in a bag. If Jenakan bought 16 bags for his friends, how many sweets will they have? _____

9. if 230 words fit on a page. How many words will fit on 15 pages?

_____

10. The area of a rectangle is length times width.
If a rectangular piece of paper has a length of 23cm and a width of 17cm, what is the area of the paper?

_____ cm²

3.3 Short division

Words for division include: divide and quotient.

Division is splitting into equal parts or groups. It is the opposite of multiplication.

So, if 386 x 83 = 32 038
then 32 038 ÷ 83 = 386
and 32 038 ÷ 386 = 83

A division problem can be shown in three different ways.

$$75 \div 3 \qquad 3\overline{)75} \qquad \frac{75}{3}$$

These three methods all mean exactly the same thing.

Unlike addition, subtraction and multiplication, in division we work from the largest place value. Divide each number with the divisor in turn.

$$\begin{array}{r}2\\4\overline{)896}\end{array}$$

Start by dividing the first digit (8) by the divisor (4). 8÷4=2.
Write the answer above the line.

$$\begin{array}{r}2\,2\\4\overline{)89\,^16}\end{array}$$

Then divide the next digit.
9÷4 = 2 with one left over.
So, write the 2 above the nine and carry the one.

$$\begin{array}{r}2\,2\,4\\4\overline{)89\,^16}\end{array}$$

Now, with the one that has been carried, the six has become 16.
16÷4 = 4.

Sometimes, the divisor does not go into a number exactly. When this happens, we can do one of two things.

- Write the number and the remainder. The remainder is the bit left over.
- We can put a decimal point and add as many zeros as we need.

$$\begin{array}{r}1\\5\overline{)9\,^476}\end{array}$$

9 ÷ 5 = 1, with four remaining.
Write 1 above the 9 and carry the 4.

$$\begin{array}{r}1\,9\\5\overline{)9\,^47\,^26}\end{array}$$

47 ÷ 5 = 9, with two remaining.
Write 9 above the 7 and carry the 2.

$$\begin{array}{r}1\,9\,5\\5\overline{)9\,^47\,^26}\end{array}$$

26 ÷ 5 = 5, with one remaining.
Write 5 above the 6 and my remainder is one.

If using remainders, my answer is now 195 r1.

However, if using decimals I continue, by writing a decimal point, top and bottom and adding a zero. Remember to carry the remainder. I can continue to add as many zeros as I need.

$$5 \overline{)9\,^47\,^26\,.\,^10} \begin{array}{c} 1\,9\,5\,. \\ \end{array}$$

Then I continue.

$$5 \overline{)9\,^47\,^26\,.\,^10} \begin{array}{c} 1\,9\,5\,.\,2 \\ \end{array}$$

Exercise 3.3

1.

$$5 \overline{)9\ 8\ 5}$$

2.

$$6 \overline{)3\ 2\ 4}$$

Calculate the following using remainders

3.

$$3 \overline{)5\ 8\ 3}$$

4.

$$8 \overline{)8\ 5\ 7\ 6}$$

5.

$$7 \overline{)6\ 7\ 4\ 3}$$

6.

$$5 \overline{)4\ 8\ 8\ 2}$$

Answer the following exactly, giving a decimal answer.

7.

$$5 \overline{)5\ 9\ 4}$$

8.

$$4 \overline{)7\ 3\ 3}$$

Give the following answers rounded to one decimal place.

(Complete to two decimal places and round.)

9.

$$6 \overline{)7\ 5\ 4}$$

10.

$$9 \overline{)4\ 9\ 0}$$

3.4 Long Division

In long division, the answer to each step is written below and then subtracted to find the remainder. The next number is brought down, to divide next.

$$\begin{array}{r} 1 \\ 5\overline{)976} \end{array}$$

5 goes into 9 once. Write 1 above the 9.

1 x 5 = 5, so write 5 below the 9. Subtract

$$\begin{array}{r} 1 \\ 5\overline{)976} \\ -5 \\ \hline 4 \end{array}$$

Bring down the next number.

$$\begin{array}{r} 1 \\ 5\overline{)976} \\ -5 \\ \hline 47 \end{array}$$

5 goes into 47, nine times. So put 9 above the 7.

$$\begin{array}{r} 19 \\ 5\overline{)976} \\ -5 \\ \hline 47 \end{array}$$

9 x 5 = 45. So write 45 under the 47 and subtract.

```
      1 9
   ┌─────────
 5 │ 9 7 6
  - 5
   ─────
     4 7
   - 4 5
     ─────
         2
```

Bring the next number down.

```
      1 9
   ┌─────────
 5 │ 9 7 6
  - 5
   ─────
     4 7
   - 4 5
     ─────
       2 6
```

5 goes into 26 five times. So put 5 above the 6

```
      1 9 5
   ┌─────────
 5 │ 9 7 6
  - 5
   ─────
     4 7
   - 4 5
     ─────
       2 6
```

5 x 5 = 25. So write 5 under the 26 and subtract.

```
      1 9 5
   ┌─────────
 5 │ 9 7 6
  - 5
   ─────
     4 7
   - 4 5
     ─────
       2 6
     - 2 5
       ─────
           1
```

There are no more digits, so put a decimal point top and bottom and add a zero to the bottom.

```
        1 9 5.
    5 | 9 7 6. 0
      - 5
        ---
        4 7
      - 4 5
        ---
          2 6
        - 2 5
          ---
            1
```

Then bring down the zero.

```
        1 9 5.
    5 | 9 7 6. 0
      - 5
        ---
        4 7
      - 4 5
        ---
          2 6
        - 2 5
          ---
            1 0
```

5 goes into 10 twice

```
      1 9 5. 2
  5 | 9 7 6. 0
    - 5
      ‾
      4 7
    - 4 5
      ‾‾‾
        2 6
      - 2 5
        ‾‾‾
          1 0
```

5 x 2 = 10. So I put that under the 10 and then subtract.

```
      1 9 5. 2
  5 | 9 7 6. 0
    - 5
      ‾
      4 7
    - 4 5
      ‾‾‾
        2 6
      - 2 5
        ‾‾‾
          1 0
        - 1 0
          ‾‾‾
            0
```

This has a remainder of zero, so my answer is 195.2.

Long division is often used with larger divisors.

Exercise 3.4

Complete the following using long division.

1.

$$11 \overline{)5\ 7\ 6\ 4}$$

2.

$$12 \overline{)7\ 8\ 4\ 8}$$

3.

$$15 \overline{)5\ 5\ 3\ 5}$$

4.

$$21 \overline{)1\ 3\ 2\ 3}$$

Answer exactly using decimals.

5.

$$32 \overline{)2\ 0\ 2\ 4}$$

6.

$$18 \overline{)8\ 1\ 9}$$

Round to one decimal place.

7.

$$13 \overline{) 4\ 9\ 2}$$

8.

$$16 \overline{) 3\ 4\ 5\ 8}$$

9.

$$22 \overline{) 7\ 8\ 6\ 9}$$

10.

$$15 \overline{) 5\ 8\ 8\ 4}$$

Exercise 3.5

You will need to use your working out book, to answer these.

1. George buys 47 boxes of mangoes costing $6 each. How much does he pay? _____

2. Petra spent $387 on books. If each book cost $3 how many books did she buy? _____

3. Lareena has all 54 books of an adventure series. If each book is worth $4, how much is the whole series worth? (assume it's worth no more for being a complete set). _____

4. A page fits 250 words of typing. How many words fit on 17 pages? _____

5. A burst water main leaks 163 litres of water per minute. How much water would be lost in three minutes? _____

6. A recipe asks for 150g of flour. If Louisa has 1250g of flour, how many batches can she make?

    _____

7. Jim runs 24 miles a day for a week (7 days). How many miles does he run in total?

    _____

8. Individual fruit juice drinks are sold in packs of three. There are 12 packs in a box and 20 boxes in a crate. How many individual fruit juice drinks are there in a crate?

    _____

9. Sweets are sold in packs of 14. How many packs does Ade need to buy if he needs 200 sweets?

    _____

10. There are 8 children at a party. For the party games each child needs 3 balloons. If there are 10 balloons in a pack, how many packs are needed?

    _____

Chapter 4: Further calculations

4.1 Addition and subtraction of negatives.

On a number line, the numbers on the left of the zero are negative and the numbers on the right are positive.

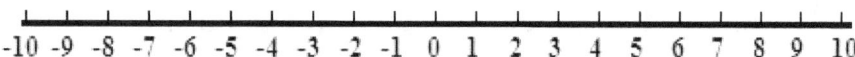

The digits get bigger as we move away from zero. A maths problem that starts with a positive number but ends with a negative one, or starts with a negative number and ends with a positive one, must be done in two steps. This is done by working out how far to zero and then how much further we need to go above or below zero.

To subtract a larger number from a smaller one. For example 4 – 9

- Determine how far to zero (in our example 4)
- Work out how many numbers left to subtract (in our example 9-4 = 5).

Therefore, in our example the answer is -5 as we need to subtract 5 from zero.

To add a number from a negative number. For example -3 + 7

- Determine how far to zero (in our example 3)
- Work out how many numbers left (in our example 7-3 = 4)

Therefore, in our example the answer is 4 as we have four numbers after zero.

To multiply a positive and a negative number, the sign becomes negative.

For example:

   2 x 4 = 8

   -2 x 4 = - 8 (the negative means we need to change the sign)

but -2 x -4 = 8 (the second negative changes the sign back to positive)

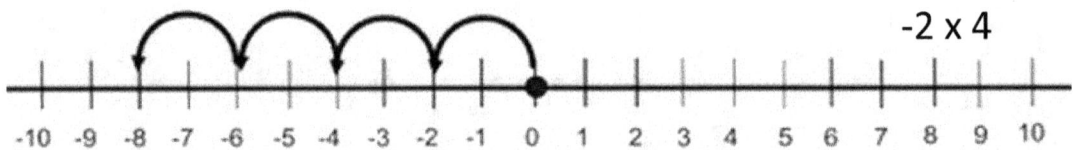

The rule of negative numbers states that two like signs make a positive and two unlike signs make a negative.

+	and	+	=	+
+	and	-	=	-
-	and	+	=	-
-	and	-	=	+

For example: 3 + (-2) is the same as 3-2
    but 3 – (-2) is the same as 3+2

Exercise 4.1

In these questions you may use the number line below.

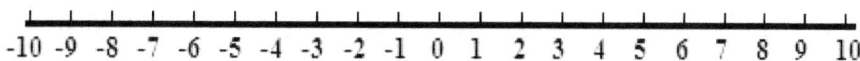

1.   5 – 9 = _____

2.   -3 + 7 = _____

3.   -23 – 4 = _____

4.   -11 + 25 = _____

5.   16 – 24 = _____

6.   -5 + 21 = _____

7.   4 x -8 = _____

8.   -3 x -6 = _____

9.   3 + (-2) = _____

10.   5 – (-4) = _____

4.2 BIMDAS

BIMDAS also called BIDMAS, BODMAS or PEMDAS refers to the order in which a maths problem should be done.

BIMDAS stands for Brackets, Indices, Multiplication and Division, Addition and Subtraction.

When doing a maths problem, the order that the parts of the problem should be calculated is:

- Brackets
- Indices (we'll look at indices later but includes square and cube numbers)
- Multiplication and division (left to right)
- Addition and subtraction (left to right)

> **Remember**
>
> B = Brackets
>
> I = Indices
>
> M = Multiplication
>
> D = Division
>
> A = Addition
>
> S = Subtraction

Example: $3 + 5 \times (9 - 7) =$

Brackets are done first. $9-7 = 2$.

$$3 + 5 \times (\underset{2}{9 - 7}) =$$

Next is the multiplication as multiplication comes before addition and subtraction.

$$3 + \underset{10}{5 \times (2)} =$$

This leaves 3+10, which is 13.

Therefore, $3 + 5 \times (9 - 7) = 13$.

Exercise 4.2

Calculate the following:

1. 5 x 3 + 2 = _____
2. 9 + 6 ÷ 3 = _____
3. 4 -2 + 6 x 3 = _____
4. 9 ÷ 3 +6 x 2 = _____
5. 3 x 2 + 12 – 8 = _____

6. (81 ÷ 9) + 2 x 4 = _____
7. 7 + 3 x 2 – 6 x 0 = _____
8. 3 x (2 + 3) – 5 = _____
9. (6 ÷ 2) x (3 + 7) = _____
10. 4 x 4 + 4 x 4 + 4 – 4 x 4 = _____

4.3 Inverse problems

Addition and subtraction are opposites.

Multiplication and division are opposites.

To go backwards in a problem, simply do the opposite.

Example: If a number is multiplied by 3 and then four was added, giving 28. What was the number?

To get 28 – 4 was added – to do the opposite we subtract 4 – giving 24
To get 24 – it was multiplied by 3 – to do the opposite we divide – giving 8

So the number was 8.

Exercise 4.3

1. I think of a number, subtract 23 and then divide by 8. The answer is 6. What number did I start with? _____

2. A number is doubled, then four is added making 26. What was the number? _____

3. A shop cost $56.20 after a $15 off voucher was applied. What was the cost of the shop before the voucher was used? _____

4. When asked for her age, Sarah said that if it was multiplied by three and one added she would be one hundred. How old is Sarah? _____

5. Christine said that she was four times her daughter's age and her daughter was double her son's age. Her son is 5.
How old is Christine? _____

6. To work out how much tax she would need to pay, Joelle subtracted 9000 and divided by five. If Joelle has to pay $2200 tax, what was her salary? _____

7. Michael thinks of a number, multiplies by six, adds 9 then divides by 3 to get 15. What was the starting number? _____

8. Veronica has a box of chocolates. She gives half to Rodney and eats 6. There are 12 chocolates left. How many chocolates were in the box? _____

9. Ezekiel spends twice as long doing his homework on Tuesday than Monday. On Wednesday he spends 50 minutes doing his homework, which is ten minutes more than Tuesday. How long did he spend doing his homework on Monday? _____

10. Sharon started to read a book on Monday. On Tuesday she read 20 more pages than on Monday. She reached page 96. How many pages did Sharon read on Monday? _____

4.4 Magic Squares

In a magic square all rows, columns and diagonals add up to the same number. To work out what a square is, first work out what the number is by adding up any complete row. Then solve any missing squares that are needed by using rows where only one number is missing.

Example: Find the value of the ?

8		
3	5	
4	?	

First work out what each row is worth by adding up the complete row.

$$8 + 3 + 4 = 15$$

Each row is worth 15.

Now solve for bottom right by using the diagonal.

$$8 + 5 = 13, \text{ the row must total 15. } 15 - 13 = 2.$$

The bottom left is 2.

8		
3	5	
4	?	2

Now solve the middle bottom. 4+2 = 6. The row must add to 15 and 15-6=9

Therefore, the missing number is 9.

Exercise 4.4

1.

	2	
?	10	
	18	4

2.

45	20	25
10		50
	?	15

3.

8	1	
	5	7
	9	?

4.

5	10	
4		8
9		?

5.

4	?	8
		1
		6

6.

12	5	10
?		11

7.

		?
7		
6	1	8

8.

4		8
		5
	?	14

9.

	?	9
12		2
		10

10.

	?	13
15	1	11

Exercise 4.5

1. The temperature in the evening is -2°C. If the temperature drops a further 5°, how cold is it overnight? _____

2. The temperature overnight drops to -3°C but over the morning rises by 10°. What temperature does it become? _____

3. Frank had -$30 in his bank account and deposited $80 into the account. What is his new bank balance? _____

4. 5 + 3 x 2 = _____

5. 7 x (2 + 3) + 2 = _____

6. I think of a number, multiply by four and add six. If the answer is 30, what number did I start with? _____

7. Tom grew 12 cm over the last year and is now 157 cm. How tall was he a year ago? _____

8. Robert gave half of his sticker collection to Shelley and then gave a further 4 away to Tom. If he still has 28 stickers, how many did he start with. _____

Work out these magic squares.

9.

	5	12
		7
?		8

10.

9		?
	6	
4		3

Puzzle Page 1

Why did the girl wear glasses when doing her maths homework?

5782 + 364 + 97 − 12 =   6231   c

1. One thousand and twelve as a number.   _____ ____

2. 926 + 318 + 14 =   _____ ____

3. 5064 − 281 =   _____ ____

4. 280 x 4 =   _____ ____

5. 30 000 ÷ 25 =   _____ ____

6. 1561 x 3 =   _____ ____

7. 0.6 x 0.7 =   _____ ____

8. 16.2 x 7 =   _____ ____

9. -25 x -8 =   _____ ____

10. 10116 ÷ 90 =   _____ ____

11. -93 + 107 =   _____ ____

12. -75 − 125 =   _____ ____

13. 210 ÷ 50 =   _____ ____

Work out what letter each of your answers represents below:

4.2	14	1012	1200	0.42	200	1258
r	p	t	m	y	n	i

6231	4683	-200	113.4	1120	4783	112.4
c	d	s	o	h	e	v

Now write the letter above the question numbers, below to answer the puzzle.

___ ___ ___ ___   ___ ___ ___ ___ ___ ___ ___
 1   4   3   7   2   5   11 13   8 10   3

___ ___ - ___ ___ ___ ___ ___ ___
 6   2   10 2   12 2   8   9

Chapter 5: More Number

5.1 Factors

A factor is a number that goes into another number with no remainder. Or put another way a factor of a number is any number that divides into it exactly.

To work out the factor of a number, start with one and work in factor pairs, working inwards to the middle.

Example: List the factors of 20

Start with 1. The factor pair of 1 is 20

$$1, \qquad\qquad 20$$

Then go to 2, 2 is a factor. The factor pair is 10.

$$1,\ 2, \qquad\qquad 10,\ 20$$

Then go to 3, 3 is not a factor so we move on to 4.
4 is a factor and the factor pair is 5.

$$1,\ 2,\ 4,\ 5,\ 10,\ 20$$

The next number is 5, so we know we have them all and can stop.

By working in like this, we know we have them all and we can work out only the larger number which we know are factors.

It is useful when working our factors to know the divisibility rules which are summarised in the table below.

Number	Divisible if	Example
2	Even – ends in 2, 4, 6, 8 or 0	72; ends in 2
3	Digits add up to 3, 6 or 9	39; 3 + 9 = 12 then 1 + 2 = 3
4	The last two digits are divisible by 4	912; 12 is divisible by 4; 12÷3
5	Ends in 0 or 5	65
6	Divisible by 2 and 3	96
7	Double and subtract the last digit from the rest of the digits. If divisible by 7, then the number is.	392; double 2 is 4, then 39-4 = 35, 35 is divisible by 7, so 392 is
8	The last three digits are divisible by 8	4120; 120 is divisible by 8
9	Digits add up to 9	153; 1 + 5 + 3 =9
10	Ends in 0	230

The prime factors of a number are the prime numbers which when multiplied together come to the number. To work out the prime factors of a number, keep dividing by the smallest possible prime number.

> Example: What are the prime factors of 20?
>
> Start with 2.   20÷2 = 10   Prime factors = 2
>
> Still divisible by 2, so divide by 2 again.
>   10÷2 = 5   Prime factors = 2 x 2
>
> 5 is a prime number so that will be our last prime factor.
>
> Therefore, the prime factors of 20 are 2 x 2 x 5 (this can also be written as $2^2 \times 5$)

Exercise 5.1

What are the factors of the following numbers?

1. 5 _____

2. 10 _____

3. 22 _____

4. 24 _____

5. 7 _____

6. 36 _____

7. 28 _____

8. 15 _____

9. 16 _____

10. 64 _____

5.2 Multiples

The multiples of a number are the numbers in the times table.

Example: What are the first six multiples of 4

    4, 8, 12, 16, 20, 24

Exercise 5.2

Write the first six multiples of the following:

1. 12 _____
2. 4 _____
3. 7 _____
4. 15 _____
5. 20 _____
6. 16 _____

Write the first four multiples of the following:

7. 17 _____
8. 25 _____
9. 14 _____
10. 36 _____

5.3. Prime numbers

A prime number is a number which only has two factors: itself and one.

One is not considered a prime number.

The prime numbers less than 20 are: 2, 3, 5, 7, 11, 13, 17, 19

A number that can be used in any multiplication fact that does not include one is not prime. These numbers are called composite numbers or rectangular numbers. Therefore, any positive number that is not one and not prime must be rectangular.

> Example: Is 27 prime or rectangular?
>
> 9 x 3 = 27
>
> therefore it must be rectangular.

Exercise 5.3

After each of these numbers write P for prime, R for rectangular or N for neither.

1. 23 _____
2. 15 _____
3. 12 _____
4. 13 _____
5. 19 _____
6. 57 _____

What are the prime factors of the following numbers?

7. 210 _____

8. 24 _____

9.   360 _____

10. 175 _____

5.4 Squares and square roots

The square of a number is the number multiplied by itself. It can be written with a "superscript 2". So, 5 squared can be written 5^2.

> Example: What is 6^2?
>
> 6^2 means 6 x 6 = 36

The square root of a number is the opposite of the square.

So, the square root of a number is what number when multiplied by itself gives this number. A square root can be written with a square root sign. So the square root of 25 can be written as $\sqrt{25}$.

> Example: What is $\sqrt{81}$?
>
> $\sqrt{81}$ = 9, because 9 x 9 = 81

Exercise 5.4

1. $3^2 = $ _____
2. $9^2 = $ _____
3. $12^2 = $ _____
4. $15^2 = $ _____
5. $13^2 = $ _____

6. $100^2 = $ _____
7. $\sqrt{9} = $ _____
8. $\sqrt{64} = $ _____
9. $\sqrt{121} = $ _____
10. $\sqrt{49} = $ _____

Exercise 5.5

What are the factors of:

1. 27 _____
2. 18 _____
3. 48 _____

Write the first four multiples of the following:

4. 8 _____
5. 6 _____
6. 13 _____

7. Is 29 rectangular or prime? _____

8. Write the answer as a number squared
   $3^2 + $ _____ $= 5^2$

9. $\sqrt{36} = $ _____

10. $25^2 = $ _____

Chapter 6: Number Relationships

6.1 Number lines

To work out the value of a position on a number line or scale, first work out what each division is worth. Then count how many divisions since the last written number.

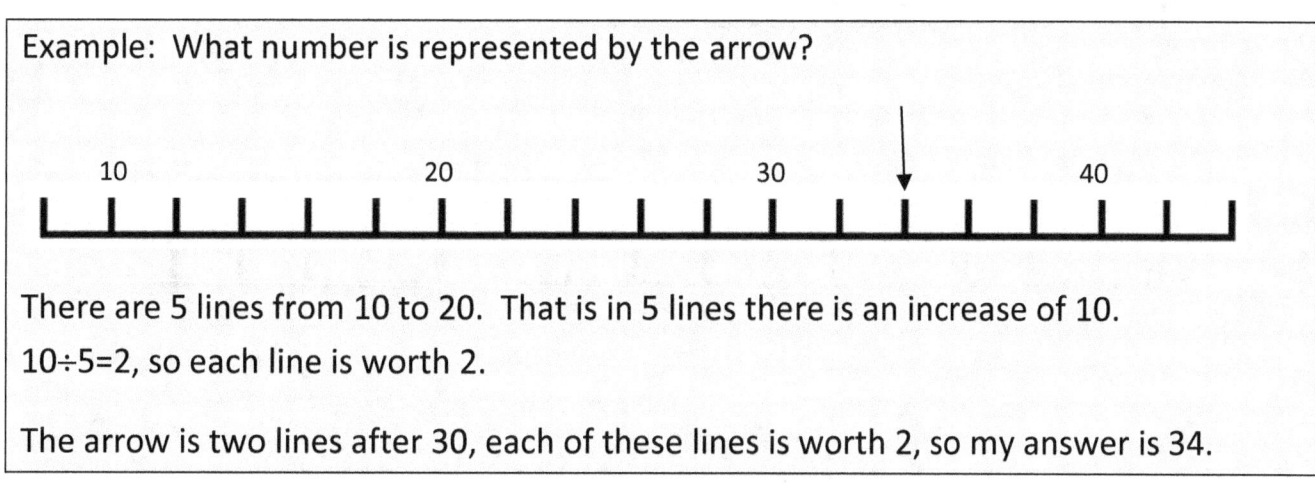

Example: What number is represented by the arrow?

There are 5 lines from 10 to 20. That is in 5 lines there is an increase of 10. 10÷5=2, so each line is worth 2.

The arrow is two lines after 30, each of these lines is worth 2, so my answer is 34.

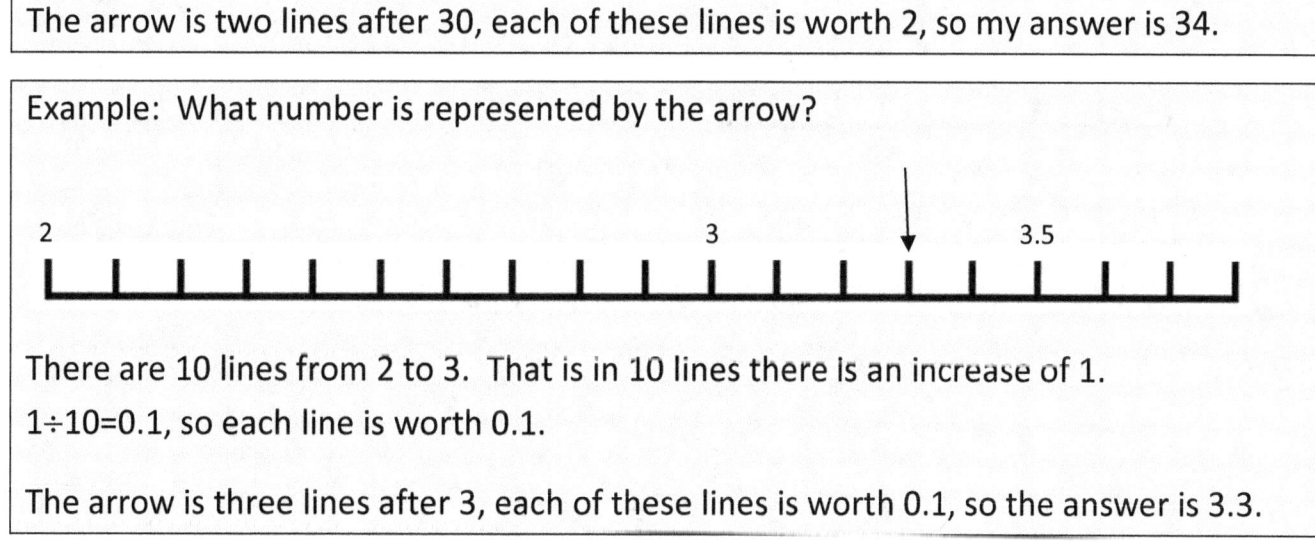

Example: What number is represented by the arrow?

There are 10 lines from 2 to 3. That is in 10 lines there is an increase of 1. 1÷10=0.1, so each line is worth 0.1.

The arrow is three lines after 3, each of these lines is worth 0.1, so the answer is 3.3.

Exercise 6.1

In the following, write the number represented by the ?.

1.
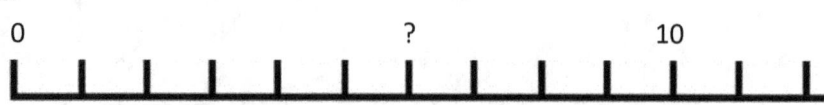

2.
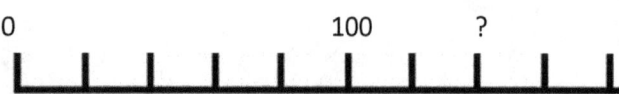

3.
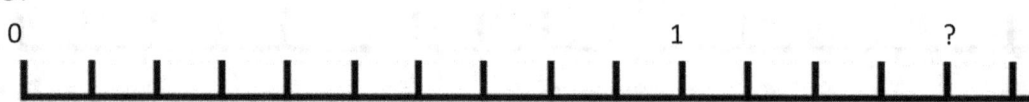

4.
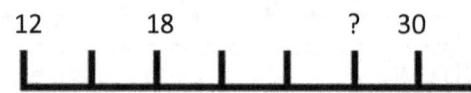

5.

6.
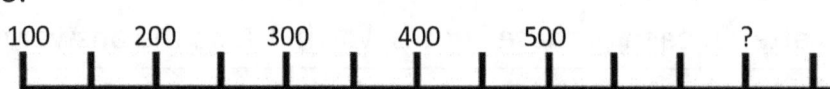

7.

8.

9 12 15 18 21   ?

9.

2.7 3.1 3.5 3.9 ?

10.

2   3   4   ?

6.2 Indices

An index is a raised number which indicates how many times the base (the big number) needs to be multiplied by itself. The plural of index is indices.

So, $2^2 = 2 \times 2 = 4$

$2^3 = 2 \times 2 \times 2 = 8$

$2^4 = 2 \times 2 \times 2 \times 2 = 16$

$2^5 = 2 \times 2 \times 2 \times 2 \times 2 = 32$

Exercise 6.2

What do the following mean? Do not calculate.

Example: 74^3 means $74 \times 74 \times 74$

1. 37^4 means _____
2. 58^5 means _____
3. 19^3 means _____

Calculate:

4. $2^3 = $ _____

5. $1^7 = $ _____

6. $4^3 = $ _____

7. $2^5 = $ _____

8. $3^5 = $ _____

9. $5^3 = $ _____

10. $10^6 = $ _____

6.3 Highest common factor (HCF)

The highest common factor of two or more numbers is the largest number that divides exactly into all numbers.

To work out the highest common factor:

- Work out the factor of all numbers
- Find the largest number that appears in all factor lists.

> Example: What is the highest common factor of 12 and 18?
>
> Factors of 12 – 1, 2, 3, 4, 6, 12
>
> Factors of 18 – 1, 2, 3, 6, 9, 18
>
> The highest common factor is 6.
>
> Note: 2 and 3 are common factors but they are not the highest common factor.

Exercise 6.3

What is the highest common factor of the following?

1. 6, 9 _____
2. 12, 15 _____
3. 35, 45 _____
4. 12, 18 _____
5. 13, 25 _____
6. 12, 24, 30 _____

Use the highest common factor to solve the following.

7. The Maths Club had a party at school. There were 20 chocolate brownies and 40 slices of pizza to be shared equally. Each student had the same number of brownies and the same number of slices of pizza with nothing left over. What is the maximum number of students who could have been at the party? _____

8. Pam has 16 red flowers and 24 yellow flowers. She wants to make bouquets with the same number of each colour flower in each bouquet. What is the greatest number of bouquets she can make? _____

9. Craig has 21 tennis balls and 14 basket balls. If he wants to divide them into identical groups without any balls left over, what is the greatest number of groups Craig can make? _____

10. Judy has 18 oranges, 24 pears and 36 bananas. She wants to make fruit baskets with the same number of each fruit in each basket. What is the greatest number of fruit baskets she can make? How many of each type of fruit will be in the baskets?
    Number of baskets: _____
    Each basket contains: ____ oranges, ____ pears and ____ bananas.

6.4 Lowest common multiple (LCM)

The lowest common multiple is the smallest number which appears in all the times tables.

> Example: What is the lowest common multiple of 6 and 9?
>
> Multiples of 6 – 6, 12, 18, 24, 30, 36, 42, 48, 54
>
> Multiples of 9 – 9, 18, 27, 36, 45, 54
>
> The lowest common multiple is 18.
>
> Note: 36 and 54 are common multiples but they are not the lowest common multiple.

There are two ways to work out the lowest common multiple:

One way to work out the lowest common denominator, is to use the prime factors for each number. To do this:

- Work out what the prime factors are for each number.
- Multiply each factor the greatest number of times it occurs in either number.

The second method is to work out the first few multiples of the largest number and then determine whether the smaller number goes into each multiple.

Exercise 6.4

What is the lowest common multiple of the following?

1. 6, 9 _____
2. 9, 21 _____
3. 5, 25 _____
4. 6, 9, 12 _____
5. 12, 18 _____
6. 4, 8, 12 _____

Use the lowest common multiple to solve the following.

7. Juliet is buying flower bulbs for her garden. She wants to buy the same number of tulips and daffodils. If tulip bulbs come in packs of 8 and daffodil bulbs come in packs of 12, what is the least number of each type of bulb Juliet will need to buy? _____

8. Two trains leave the station at the same time. If one train departs every 8 minutes and one every 12 minutes. How many minutes before they both leave at the same time again? _____

9. Pencils come in packs of 12. Erasers come in packs of 5. Joshua wants to purchase the smallest number of pencils and erasers so that he will have exactly 1 eraser per pencil. How many packs of pencils should Joshua buy? _____

10. Ha and Tuan are shelving books at a public library. Ha shelves 12 books at a time, whereas Tuan shelves 15 at a time. If they end up shelving the same number of books, what is the smallest number of books each could have shelved? _____

Exercise 6.5

What does the question mark represent in the following number lines?

1.
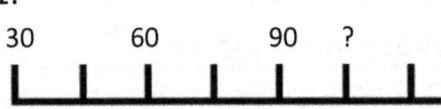

2.
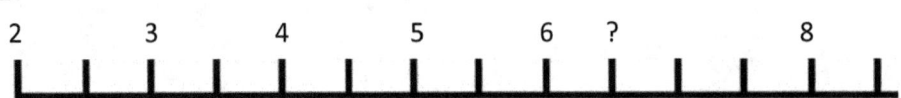

3.
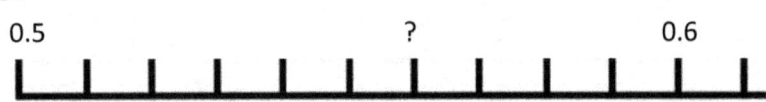

Calculate:

4. $2^8 =$ _____

5. $3^4 =$ _____

What is the highest common factor of:

6. 26, 39 _____

7. 54, 108, 36 _____

What is the lowest common multiple of:

8. 3, 5, 6 _____

9. Boxes that are 12 inches tall are being stacked next to boxes that are 18 inches tall. What is the shortest height at which the two stacks will be the same height? _____

10. Autumn is making shoeboxes for a Christmas charity appeal. If she has 12 notepads, 18 bars of soap and 36 packs of colouring pencils. What is the largest number of boxes she can make if each item is evenly distributed among the boxes, with none left over? _____

Chapter 7: Fractions

7.1 Simple fractions

A fraction is part of a whole. A fraction consists of two numbers:

- The numerator is the top number and represents the number of parts selected or being referred to.
- The denominator is the bottom number and represents the total number of parts that 1 whole is divided into.

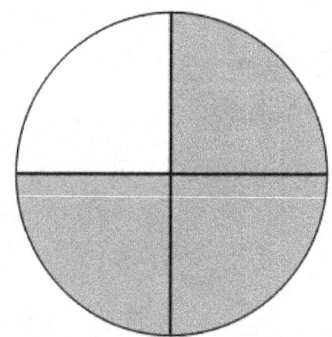

So, in this circle, there are 3 sections shaded out of 4, therefore my fraction is $\frac{3}{4}$.

A fraction line is the same as a divide by sign.

> Example: $\frac{3}{4}$ is the same as 3 ÷ 4.

Exercise 7.1

What fractions are represented below?

1. 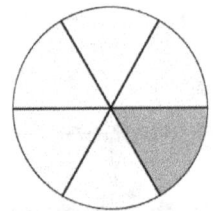 Fraction shaded: Fraction unshaded:

2. 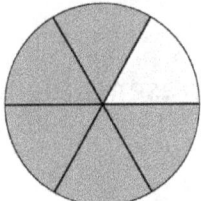 Fraction shaded: Fraction unshaded:

3. 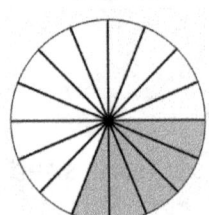 Fraction shaded: Fraction unshaded:

4. 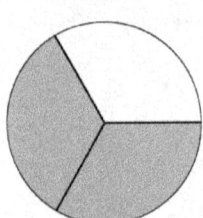 Fraction shaded: Fraction unshaded:

5. 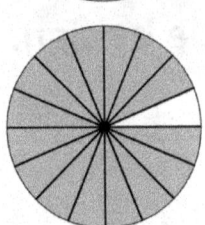 Fraction shaded: Fraction unshaded:

Write the following fractions as divisions (eg. $\frac{3}{4} = 3 \div 4$)

6. $\frac{7}{9}$ = _____

7. $\frac{11}{17}$ = _____

8. $\frac{2}{3}$ = _____

9. $\frac{5}{7}$ = _____

10. $\frac{5}{12}$ = _____

7.2 Equivalent fractions

If I have a cake and cut it into four pieces and have one, or cut it into eight pieces and have two, I am eating the same amount of cake. We call these equivalent fractions.

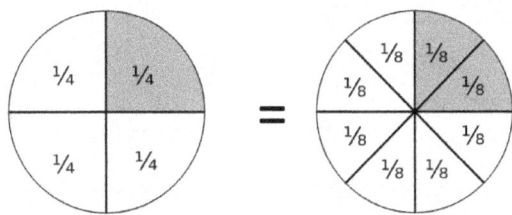

This table shows some more equivalent fractions.

1											
$\frac{1}{2}$						$\frac{1}{2}$					
$\frac{1}{3}$				$\frac{1}{3}$				$\frac{1}{3}$			
$\frac{1}{4}$			$\frac{1}{4}$			$\frac{1}{4}$			$\frac{1}{4}$		
$\frac{1}{5}$		$\frac{1}{5}$		$\frac{1}{5}$		$\frac{1}{5}$			$\frac{1}{5}$		
$\frac{1}{6}$		$\frac{1}{6}$		$\frac{1}{6}$		$\frac{1}{6}$		$\frac{1}{6}$		$\frac{1}{6}$	
$\frac{1}{7}$	$\frac{1}{7}$		$\frac{1}{7}$		$\frac{1}{7}$		$\frac{1}{7}$		$\frac{1}{7}$		$\frac{1}{7}$
$\frac{1}{8}$	$\frac{1}{8}$		$\frac{1}{8}$	$\frac{1}{8}$		$\frac{1}{8}$	$\frac{1}{8}$		$\frac{1}{8}$		$\frac{1}{8}$
$\frac{1}{9}$	$\frac{1}{9}$	$\frac{1}{9}$	$\frac{1}{9}$	$\frac{1}{9}$	$\frac{1}{9}$	$\frac{1}{9}$	$\frac{1}{9}$		$\frac{1}{9}$		
$\frac{1}{10}$	$\frac{1}{10}$	$\frac{1}{10}$	$\frac{1}{10}$	$\frac{1}{10}$	$\frac{1}{10}$	$\frac{1}{10}$	$\frac{1}{10}$	$\frac{1}{10}$	$\frac{1}{10}$		
$\frac{1}{11}$	$\frac{1}{11}$	$\frac{1}{11}$	$\frac{1}{11}$	$\frac{1}{11}$	$\frac{1}{11}$	$\frac{1}{11}$	$\frac{1}{11}$	$\frac{1}{11}$	$\frac{1}{11}$	$\frac{1}{11}$	
$\frac{1}{12}$	$\frac{1}{12}$	$\frac{1}{12}$	$\frac{1}{12}$	$\frac{1}{12}$	$\frac{1}{12}$	$\frac{1}{12}$	$\frac{1}{12}$	$\frac{1}{12}$	$\frac{1}{12}$	$\frac{1}{12}$	$\frac{1}{12}$

To work out equivalent fractions, simply multiply or divide both the top and bottom of the fraction by the same number.

Remember: We can multiply or divide the top number by anything as long as to the bottom number we do *exactly the same*.

$$\text{So: } \frac{1 \text{ x2}}{4 \text{ x2}} = \frac{2}{8}$$

Exercise 7.2a
Work out the missing number, in the questions below:

1. $\frac{3}{10} = \frac{}{50}$

2. $\frac{27}{30} = \frac{}{10}$

3. $\frac{4}{10} = \frac{}{5}$

4. $\frac{1}{4} = \frac{}{12} = \frac{7}{}$

5. $\frac{3}{8} = \frac{}{40} = \frac{18}{}$

6. $\frac{6}{42} = \frac{1}{}$

7. $\frac{21}{57} = \frac{7}{}$

8. $\frac{5}{8} = \frac{25}{} = \frac{}{160}$

9. $\frac{11}{44} = \frac{33}{} = \frac{1}{}$

10. $\frac{5}{6} = \frac{}{30} = \frac{150}{}$

One as a fraction
A fraction that has the same numerator and denominator has a value of one.

For example: If I have a pizza, that is cut into 8 pieces and I have all 8 pieces, then I have $\frac{8}{8}$ of the pizza or 1 pizza. Therefore, $\frac{8}{8} = 1$.

Exercise 7.2b

Complete the following fractions to equal one.

1. $\dfrac{}{10} = 1$

2. $\dfrac{27}{} = 1$

3. $\dfrac{4}{} = 1$

4. $1 = \dfrac{7}{}$

5. $1 = \dfrac{18}{}$

6. $\dfrac{}{42} = 1$

7. $\dfrac{}{56} = 1$

8. $1 = \dfrac{}{160}$

9. $1 = \dfrac{}{33}$

10. $\dfrac{}{30} = 1$

7.3 Simplifying fractions

Fractions can be simplified if the numerator and denominator have a common factor. This process is called cancelling. When answering with fractions you should *always* write the fraction in the simplest form.

$$\text{So: } \dfrac{27 \div 3}{30 \div 3} = \dfrac{9}{10}$$

Exercise 7.3

Simplify the following fractions.

1. $\dfrac{8}{12} = \dfrac{}{}$

2. $\dfrac{27}{81} = \dfrac{}{}$

3. $\dfrac{14}{21} = \dfrac{}{}$

4. $\dfrac{8}{24} = \dfrac{}{}$

5. $\dfrac{15}{35} = \dfrac{}{}$

6. $\dfrac{28}{48} = \dfrac{}{}$

7. $\dfrac{22}{99} = \dfrac{}{}$

8. $\dfrac{24}{96} = \dfrac{}{}$

9. $\dfrac{56}{80} = \dfrac{}{}$

10. $\dfrac{64}{96} = \dfrac{}{}$

7.4 Improper fractions and mixed numbers

A proper fraction is one where the numerator is smaller than the denominator.

For example $\frac{2}{3}$.

An improper fraction is one where the numerator is larger than the denominator.

For example $\frac{3}{2}$.

A mixed number consists of a whole number and a fraction.

For example $4\frac{2}{3}$.

While a proper fraction has a value less than one, both improper fractions and mixed numbers are worth more than one.

If I have the following amount of apple pie left

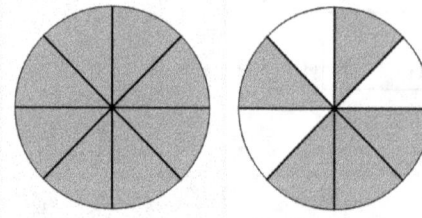

Then each apple pie is divided into 8 pieces, so my denominator is 8.

There are 13 pieces left, so the improper fraction is $\frac{13}{8}$.

Another way of looking at it is, there is one pie and five pieces remaining.

This gives the mixed number $1\frac{5}{8}$.

Exercise 7.4a

What improper fraction and mixed number are represented by the following pictures?

(Remember fractions must always be written in their simplest form).

1.		Improper fraction =	Mixed number =
2.		Improper fraction =	Mixed number =
3.		Improper fraction =	Mixed number =
4.		Improper fraction =	Mixed number =
5.		Improper fraction =	Mixed number =
6.		Improper fraction =	Mixed number =
7.		Improper fraction =	Mixed number =
8.		Improper fraction =	Mixed number =
9.		Improper fraction =	Mixed number =
10.		Improper fraction =	Mixed number =

Converting improper fractions into mixed numbers

To convert an improper fraction into a whole number, we need to work out how many whole numbers the fraction contains and then write the remainder as the fraction part.

> For example: If a recipe asks for ¼ cup of oats per person. If we are cooking for 11 people, that would be $\frac{11}{4}$ cups. First, we need to work out how many full cups there are. One cup contains 4 quarters (ie. $1=\frac{4}{4}$).
>
> There are two lots of four in 11. That is $\frac{8}{4} = 2$, so we have two whole. There are three quarters left over, so our mixed number is $2\frac{3}{4}$.

This can be simplified to the following steps.

1. Divide the numerator by the denominator (top number ÷ bottom number)
2. Write down the whole number answer
3. Write down any remainder above the same denominator.

> Example: Convert $\frac{14}{4}$ to a mixed number.
>
> Divide 14÷4 = 3 with a remainder of 2
>
> Write down the 3 and then write down the remainder (2) over the denominator (4)
>
> $$3\frac{2}{4}$$
>
> I then need to simplify as fractions are always written in their lowest terms.
>
> So my answer is $3\frac{1}{2}$

Converting mixed number to improper fractions

To convert a mixed number to an improper fraction we need to work out what the whole number would be as a fraction with the same denominator as the fraction part and then add the two fractions together.

> Example: Convert $4\frac{2}{3}$ to an improper fraction.
>
> As the denominator is 3, every whole number has 3 parts, so $4 = \frac{12}{3}$.
>
> Then add the 12 parts of the whole number to the 2 parts of the original fraction.
>
> This gives 12 + 2 = 14 parts.
>
> The denominator stays the same so the improper fraction is $\frac{14}{3}$.

This can be simplified to the following steps.

1. Multiply the whole number part by the fraction's denominator.
2. Add that to the numerator.
3. Write the result on top of the denominator.

> Example: Write $7\frac{3}{5}$ as an improper fraction.
>
> Multiply 7 x 5 = 35
>
> Add this to the numerator (3). 35 + 3 = 38
>
> Write this over the denominator (5).
>
> This gives the improper fraction, $\frac{38}{5}$.

Exercise 7.4b

Convert the following improper fractions to mixed numbers.

1. $\dfrac{27}{12} =$

2. $\dfrac{27}{5} =$

3. $\dfrac{14}{9} =$

4. $\dfrac{33}{24} =$

5. $\dfrac{77}{35} =$

Convert the following mixed numbers to improper fractions.

6. $3\dfrac{2}{5} =$

7. $1\dfrac{11}{13} =$

8. $3\dfrac{19}{20} =$

9. $8\dfrac{4}{7} =$

10. $15\dfrac{5}{6} =$

7.5 Lowest Common Denominator (LCD)

The lowest common denominator (also called the least common denominator) is the smallest denominator that two of more fractions can use to write equivalent fractions. That is it is the smallest number that can be divided exactly by all the numbers below the lines in a group of two or more fractions.

> For example: What is the lowest common denominator of $\dfrac{1}{6}, \dfrac{5}{8}$?
>
> The LCD is 24, as the fractions can be written as $\dfrac{4}{24}$ and $\dfrac{15}{24}$.

The lowest common denominator is the same as the lowest common multiple of the denominators.

As mentioned in section 6.4, one way to work out the lowest common denominator, is to use the prime factors for each number. To do this:

1. Work out what the prime factors are for each number.
2. Multiply each factor the greatest number of times it occurs in either number.

> For example: Work out the lowest common denominator of 12 and 20.
>
> The prime factors of 12 are 2, 2, 3
> The prime factors of 20 are 2, 2, 5
> So the lowest common denominator will be 2x2x3x5 = 60

Exercise 7.5

Write the lowest common denominator for the fractions below:

1. $\frac{2}{3}$ and $\frac{5}{9}$ _____

2. $\frac{7}{12}$ and $\frac{4}{9}$ _____

3. $\frac{2}{15}$ and $\frac{13}{20}$ _____

4. $\frac{3}{4}$ and $\frac{5}{6}$ _____

5. $\frac{7}{11}$ and $\frac{3}{5}$ _____

Write the following fractions as equivalent fractions using their lowest common denominator.

6. $\frac{7}{10}$ and $\frac{5}{15}$ _____

7. $\frac{1}{3}$ and $\frac{4}{9}$ _____

8. $\frac{5}{6}$ and $\frac{3}{60}$ _____

9. $\frac{5}{8}$ and $\frac{5}{12}$ _____

10. $\frac{7}{27}$ and $\frac{5}{6}$ _____

Chapter 8: Calculating Fractions

8.1 Adding fractions

To add fractions the fractions must have the same denominator. Then the numerators are added.

> For example: A pizza is cut into eight pieces. If Cameron has two pieces of pizza and Ronald has three pieces. How much pizza has been eaten.
> Cameron has $\frac{2}{8}$ of the pizza and Ronald has $\frac{3}{8}$. Altogether, they have eaten five pieces of pizza out of eight or $\frac{5}{8}$.
>
>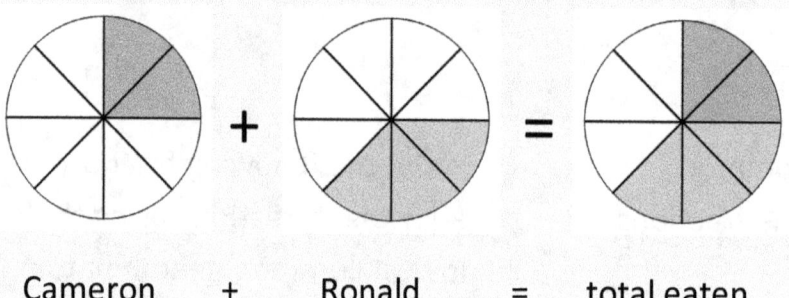
>
> Cameron   +   Ronald   =   total eaten
>
> The top numbers are added because the number of pieces referred to increases, but the bottom number stays the same as the pizza is still cut into eight pieces.
>
> So $\frac{2}{8} + \frac{3}{8} = \frac{5}{8}$

This can be simplified to the following steps.

1. If the fractions have different denominators, put them over the lowest common denominator.
2. Add the top numbers (numerators).
3. Put the answer over the denominator.
4. Simplify (if needed).

Exercise 8.1

Add the following fractions.

(Remember to simplify and convert any improper fractions to mixed numbers).

1. $\frac{2}{7} + \frac{4}{7} =$

2. $\frac{3}{11} + \frac{2}{11} =$

3. $\frac{4}{9} + \frac{2}{9} =$

4. $\frac{2}{3} + \frac{1}{2} =$

5. $\frac{3}{7} + \frac{4}{7} =$

6. $\frac{3}{5} + \frac{1}{4} =$

7. $\frac{1}{6} + \frac{4}{7} =$

8. $\frac{1}{6} + \frac{2}{9} =$

9. $\frac{3}{4} + \frac{4}{5} =$

10. $\frac{8}{9} + \frac{7}{12} =$

8.2 Subtracting Fractions

Subtracting fractions is very similar to adding fractions.

The steps for subtracting fractions are:
1. If the fractions have different denominators, put them over the lowest common denominator.
2. Subtract the top numbers (numerators).
3. Put the answer over the denominator.
4. Simplify (if needed).

For example: $\frac{3}{4} - \frac{1}{3} =$

Different denominators, so need to convert to the lowest common denominator first.

Lowest common denominator of 4 and 3 is 12.

So, question becomes $\frac{9}{12} - \frac{4}{12} =$

Subtract the numerators, 9-4 = 5

Put the result over the denominator and the answer is $\frac{5}{12}$.

Exercise 8.2

Subtract the following fractions.
(Remember to simplify and convert any improper fractions to mixed numbers).

1. $\frac{6}{7} - \frac{4}{7} =$

2. $\frac{9}{11} - \frac{2}{11} =$

3. $\frac{4}{9} - \frac{1}{9} =$

4. $\frac{2}{3} - \frac{1}{2} =$

5. $\frac{3}{5} - \frac{1}{4} =$

6. $\frac{7}{8} - \frac{5}{6} =$

7. $\frac{4}{9} - \frac{4}{12} =$

8. $\frac{1}{6} - \frac{1}{9} =$

9. $\frac{3}{4} - \frac{2}{5} =$

10. $\frac{8}{9} - \frac{7}{12} =$

8.3 Multiplying Fractions

To multiply fractions, it is easiest to cancel first. To cancel we can divide anything into the top as long as to the bottom we do exactly the same. This is called cross-cancelling.

> For example: $\frac{5}{8} \times \frac{4}{5} =$
>
> Four goes into both top and bottom. Four goes into 8 twice and 4 once. This gives us:
>
> $$\frac{5}{\cancel{8}_2} \times \frac{\cancel{4}^1}{5} =$$
>
> The two fives can also be cancelled, each going into 5 once. This gives us:
>
> $$\frac{\cancel{5}^1}{\cancel{8}_2} \times \frac{\cancel{4}^1}{\cancel{5}_1} =$$

> Then multiply across, top and bottom. 1x1=1 and 2x1=2,
> so $\frac{5}{8} \times \frac{4}{5} = \frac{1}{2}$.

The steps to multiplying fractions are:

1. Cross cancel (if possible)
2. Multiply the numerators (top numbers)
3. Multiply the denominators (bottom numbers)

If you are multiplying a fraction by a whole number, simply write the whole number with a denominator of 1.

> For example: $3 \times \frac{2}{5} =$ is the same as $\frac{3}{1} \times \frac{2}{5} =$

Exercise 8.3
Multiply the following fractions

1. $\frac{5}{8} \times \frac{1}{3} =$

2. $\frac{2}{5} \times \frac{4}{5} =$

3. $\frac{7}{11} \times \frac{2}{3} =$

4. $\frac{6}{7} \times \frac{14}{23} =$

5. $2 \times \frac{3}{4} =$

6. $\frac{3}{8} \times \frac{4}{9} =$

7. $\frac{5}{14} \times \frac{21}{30} =$

9. $\frac{2}{3} \times \frac{9}{16} =$

8. $\frac{12}{17} \times \frac{34}{40} =$

10. $\frac{11}{15} \times \frac{5}{9} =$

8.4 Dividing Fractions

To divide fractions, turn the second fraction upside down and then multiply.

For example: $\frac{3}{4} \div \frac{9}{16} =$

First turn second fraction upside down and change ÷ into x.

$$\frac{3}{4} \times \frac{16}{9} =$$

Then cross-cancel.

$$\frac{\cancel{3}^1}{\cancel{4}_1} \times \frac{\cancel{16}^4}{\cancel{9}_3} =$$

Then multiply across.

$$\frac{1}{1} \times \frac{4}{3} = \frac{4}{3} = 1\frac{1}{3}$$

Exercise 8.4

Divide the following fractions.

1. $\dfrac{2}{5} \div \dfrac{4}{5} =$

2. $\dfrac{8}{9} \div \dfrac{2}{3} =$

3. $\dfrac{3}{4} \div \dfrac{4}{8} =$

4. $\dfrac{3}{4} \div \dfrac{1}{2} =$

5. $\dfrac{6}{7} \div \dfrac{12}{21} =$

6. $\dfrac{5}{14} \div \dfrac{5}{7} =$

7. $\dfrac{4}{9} \div \dfrac{24}{27} =$

8. $\dfrac{6}{15} \div \dfrac{3}{5} =$

9. $\dfrac{2}{6} \div \dfrac{8}{11} =$

10. $\dfrac{3}{4} \div 5 =$

8.5 Calculating with mixed numbers

Adding and subtracting mixed numbers.

To add and subtract is the same as adding or subtracting other numbers. We always start by adding from the smallest place value and if there are fractions this is the fraction part.

When adding or subtracting fractions be careful when borrowing or carrying. Remember a one in the units place is worth the same value as the denominator in the fractions place.

Example: $7\frac{1}{3} - 2\frac{3}{4}$

To add or subtract denominators must be the same. LCD is 12. $7\frac{4}{12} - 2\frac{9}{12}$.

Write in columns

$$\begin{array}{r} 7\frac{4}{12} \\ - 2\frac{9}{12} \\ \hline \end{array}$$

4-9 I can't do, so I need to borrow from the next place value which is the 7. So the 7 becomes 6. The 1 in the units is worth $\frac{12}{12}$ so I add 12 to my numerator (4), making 16.

$$\begin{array}{r} {}^{6}\cancel{7}\;\frac{{}^{+12}4}{12} \\ - 2\;\frac{9}{12} \\ \hline \end{array}$$

This is the same as

$$\begin{array}{r} {}^{6}\cancel{7}\;\frac{{}^{16}\cancel{4}}{12} \\ - 2\;\frac{9}{12} \\ \hline \end{array}$$

Then subtract the numerators. 16-9=7. The denominator stays the same.

$$\begin{array}{r} {}^{6}\cancel{7}\;\frac{{}^{16}\cancel{4}}{12} \\ - 2\;\frac{9}{12} \\ \hline \frac{7}{12} \end{array}$$

Then, continue with the other place values.

$$\begin{array}{r} {}^{6}\cancel{7}\;\frac{{}^{16}\cancel{4}}{12} \\ - 2\;\frac{9}{12} \\ \hline 4\;\frac{7}{12} \end{array}$$

Exercise 8.5a

1. $1\frac{1}{6} + 2\frac{4}{6} =$

2. $3\frac{1}{4} + 2\frac{3}{4} =$

3. $4\frac{2}{11} + 3\frac{7}{11} =$

4. $3\frac{4}{5} + 2\frac{3}{5} =$

5. $6\frac{5}{6} + 3\frac{13}{15} =$

6. $7\frac{5}{6} - 2\frac{1}{6} =$

7. $9\frac{7}{9} - 5\frac{2}{9} =$

8. $3\frac{3}{7} - 1\frac{4}{7} =$

9. $4\frac{1}{8} - \frac{4}{6} =$

10. $2\frac{1}{5} - 1\frac{4}{15} =$

<u>Multiplication and division of mixed numbers</u>

To multiply or divide mixed numbers, turn the mixed number into an improper fraction and calculate as before.

> **Example:** $3\frac{1}{2} \div 1\frac{1}{4}$
>
> First convert into improper fractions. $\frac{7}{2} \div \frac{5}{4}$
>
> Then proceed as before.
> It's division, so turn the second fraction upside down and change the ÷ to a x.
>
> $\frac{7}{2} \times \frac{4}{5}$
>
> Cross cancel, then multiply across
>
> $\frac{7}{\cancel{2}_1} \times \frac{\cancel{4}^2}{5} = \frac{14}{5}$
>
> Then convert to a mixed number
>
> $\frac{14}{5} = 2\frac{4}{5}$

Exercise 8.5b

1. $\frac{2}{3} \times 3\frac{1}{4} =$

2. $4\frac{2}{7} \times 1\frac{3}{4} =$

3. $8\frac{1}{3} \times 2\frac{2}{5} =$

4. $5\frac{2}{8} \times 3\frac{3}{7} =$

5. $7\frac{1}{9} \times 3 =$

6. $5 \div \frac{5}{6} =$

7. $6\frac{1}{2} \div 3\frac{1}{4} =$

8. $1\frac{1}{3} \div 5\frac{1}{3} =$

9. $\frac{7}{8} \div 4\frac{2}{6} =$

10. $3\frac{2}{3} \div 11 =$

Chapter 9: Solving Fraction Problems

9.1 Complementary fractions

Complementary fractions are two or more fractions that add up to one. A fraction tells us how many parts are being referred to and we can work out the complement; how many parts are not being referred to in that given fraction. The complement of the given fraction can be worked out by taking the fraction away from 1. Working out the complementary fraction is very useful in solving problems involving fractions.

> Example: If Georgia gives Tom $\frac{1}{6}$ of her stamp collection, what fraction of her original collection does she have left?
>
> Remember $1 = \frac{6}{6}$
>
> So $\frac{6}{6} - \frac{1}{6} = \frac{5}{6}$
>
> She has $\frac{5}{6}$ of her stamp collection left.

Exercise 9.1

1. $\frac{3}{8}$ of the children in a class are boys. What fraction of the class are girls? _____

2. In a class $\frac{7}{8}$ of the children enjoy computer games. What fraction does not enjoy playing computer games? _____

3. On an outing, $\frac{2}{7}$ of the children chose berry flavoured ice-cream. What fraction chose a different flavour? _____

4. Simon went through his toys and gave away $\frac{3}{5}$ to a charity shop as he had grown out of them. What fraction of his toys did he keep?
_____

5. In a sale, a shop had all goods $\frac{1}{4}$ off. How much did customers still need to pay? _____

6. Marina ate $\frac{2}{5}$ of a chocolate bar. What fraction did she keep for later? _____

7. In a school magazine $\frac{2}{9}$ of the pages contained an error. What fraction of the pages were error free? _____

8. 42 children in year six took part in a netball tournament. The rest watched. If $\frac{3}{7}$ of the students played, what fraction of the students watched? _____

9. The year five students went to the library. Everyone had to take out a book. $\frac{8}{11}$ of the children took out a fiction book. What fraction of the children took out a non-fiction book? _____

10. Niall sold chocolate, strawberry and vanilla milkshakes. If $\frac{1}{12}$ of his customers bought vanilla and $\frac{4}{12}$ bought strawberry, what fraction of customers bought a chocolate flavoured milkshake? _____

9.2 Fraction of a number

The word "of" means times.

So $\frac{3}{4}$ of 200 is the same as $\frac{3}{4} \times 200$.

There are two methods to work out the fraction of a number.
1. Work out the amount of one part, then multiply the answer by the number of parts.
2. Replace "of" with times (x), cross cancel, then multiply across.

Example: $\frac{3}{4}$ of 200

Method 1

Work out $\frac{1}{4}$ (divide by 4)

$\frac{1}{4}$ is 50

Then multiply by the numerator (3)

$\frac{3}{4}$ is 50 x 3 = 150

Method 2

$\frac{3}{4}$ of 200   change "of" to "x"

$\frac{3}{4} \times \frac{200}{1}$

Cross cancel and multiply across

$\frac{3}{\cancel{4}_1} \times \frac{\cancel{200}^{50}}{1} = \frac{150}{1} = 150$

Exercise 9.2

1. $\frac{4}{5}$ of 35 =

2. $\frac{2}{7}$ of 42 =

3. $\frac{5}{6}$ of 480 =

4. $\frac{3}{4}$ of 60 =

5. $\frac{2}{5}$ of 920 =

6. $\frac{3}{8}$ of 960 =

7. There are 32 children in a class. If $\frac{3}{8}$ of the class are girls, how many girls are in the class? _____

8. A tuition centre has 1200 students. If last year $\frac{4}{5}$ of the students attending the tuition centre passed the exam, how many students did not pass? _____

9. Douglas has 70 stickers. He gives $\frac{2}{5}$ of them to Clyde. How many stickers does Clyde receive? _____

10. 72 children are going on a school trip. The school has two coaches for the trip. If $\frac{2}{3}$ of the children travel on the big coach, how many children will travel on the small coach?

_____

9.3 Working out the whole or original amount

If we have a fraction we can also work out the full amount.

Again there are two methods.

1. Work out one part (divide by numerator), then multiply by total number of parts.
2. Divide.

> Example: If $\frac{3}{5}$ is 90, what is the whole.
>
> **Method 1**
>
> $\frac{3}{5}$ is 90 (so to work out $\frac{1}{5}$; ÷ 3).
>
> $\frac{1}{5} = 30$
>
> $\frac{5}{5} = 150$
>
> So, the whole would be 150.
>
> **Method 2**
>
> $\frac{3}{5}$ is 90
>
> $\frac{90}{1} \div \frac{3}{5}$
>
> $\frac{90}{1} \times \frac{5}{3}$
>
> $\frac{\cancel{90}^{30}}{1} \times \frac{5}{\cancel{3}_1} = 150$

Exercise 9.3

What would be the whole if:

1. $\frac{2}{5}$ is 14 _____

2. $\frac{4}{9}$ is 36 _____

3. $\frac{5}{7}$ is 40 _____

4. $\frac{3}{8}$ is 36 _____

5. $\frac{5}{6}$ is 90 _____

6. $\frac{5}{8}$ is 125 _____

7. In a class, $\frac{3}{5}$ of the students are boys. If there are 21 boys, how many students are there in the whole class? _____

8. Eloise gave $\frac{2}{7}$ of her sticker collection to Amna. If Amna was given 12 stickers, how many stickers did Eloise have before she gave any away? _____

9. Aiden made a batch of muffins and sold 8 of them at a fair. If he sold $\frac{4}{9}$ of them, how many did he bake? _____

10. On a school trip to a theme park $\frac{7}{9}$ of the children chose to go on the "Screeching Dragon." If 12 children chose not to go, how many children were on the trip altogether? _____

9.4 Fraction Problems

Fraction problems can normally be solved by following these easy steps.

1. Work out what fraction you know (it may be the complementary fraction).
2. Work out what fraction you need to know.
3. Calculate what one part is.
4. Multiply to get the required number of parts.

Example: Kirsty has read $\frac{5}{8}$ of her book. She has 81 pages still to read. How many pages does the book have altogether?

(If Kirsty has read $\frac{5}{8}$ of the book she has $\frac{3}{8}$ still to read.)

$\frac{3}{8}$ is 81 pages   (If 3 parts is 81 then 1 part is 81 ÷ 3)

$\frac{1}{8}$ is 27 pages

$\frac{8}{8}$ is 27 x 8 = 216 pages.

Exercise 9.4

1. 45 children in year six took part in a netball tournament. The rest watched. If $\frac{3}{7}$ of the students played, how many students watched?

   _____

2. At a party $\frac{2}{9}$ of those present are children. If there are 49 adults at the party how many children are there?

   _____

3. A school magazine has 64 pages. If there are errors on $\frac{3}{8}$ of the pages, how many pages are error free?

   _____

4. On a test Sanjoy got $\frac{3}{10}$ of the questions wrong. If he answered 28 questions correctly, how many questions were on the test paper?

　　　　　_____

5. Annil had a box containing 24 chocolates. If he gives $\frac{1}{4}$ of them to Francis, how many chocolates does he have left?

　　　　　_____

6. Isabella had $\frac{1}{4}$ of the pizza and Anabelle ate $\frac{1}{8}$. There are 10 pieces of pizza left. How many slices did the pizza have?

　　　　　_____

7. In a woodland $\frac{3}{5}$ of the trees are beech trees and the rest are oak trees. If there are 120 oak trees, how many beech trees are there?

　　　　　_____

8. In a class of 35 students $\frac{4}{7}$ are boys. How many girls are in the class?

　　　　　_____

9. Akhil gave 12 stickers to Isaac and had $\frac{5}{6}$ of his stickers remaining. If he then gives a further 6 stickers away to Albin, how many stickers does he have left?

　　　　　_____

10. A bookshop has a deal of $\frac{1}{3}$ off, if five or more books are bought at the same time. If Gale buys 5 books for a total of $9.
How much would the books have cost if they had been bought individually?

　　　　　_____

Chapter 10: Percentages

10.1 Percentages

Percentage means out of 100.

One hundred percent is one or the whole.

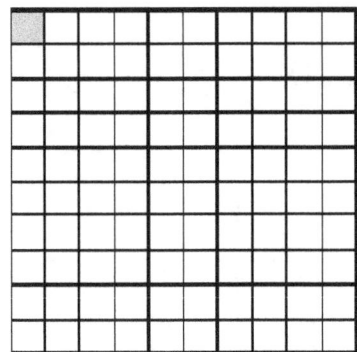 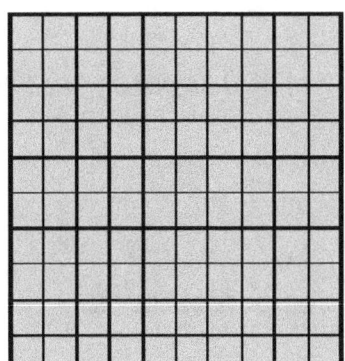

1 out of 100 squares are shaded.   100 out of 100 squares are shaded
$\frac{1}{100} = 1\%$ of squares are shaded. $\frac{100}{100} = 100\%$ of squares are shaded.

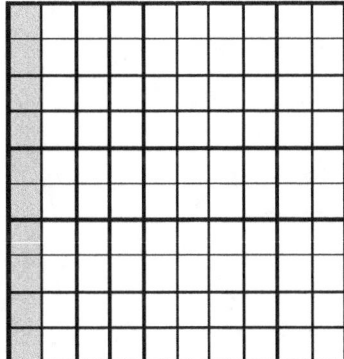 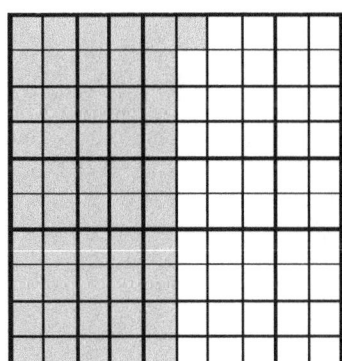

10 out of 100 squares are shaded.   51 out of 100 squares are shaded
$\frac{10}{100} = 10\%$ of squares are shaded. $\frac{51}{100} = 51\%$ of squares are shaded.

Percentages can be greater than 100%

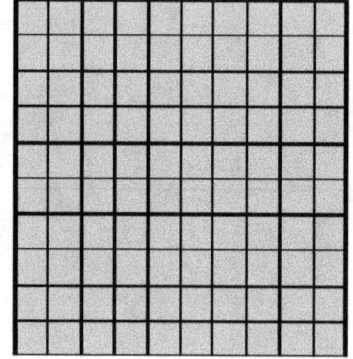 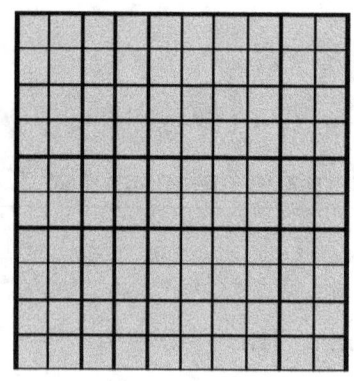 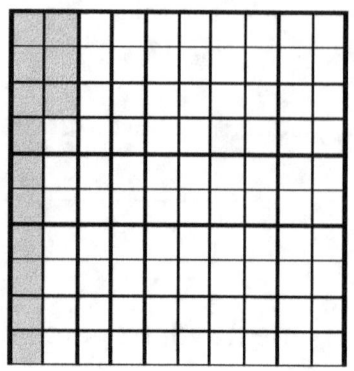

2 whole and 13 out of 100 squares are shaded = $2\frac{13}{100}$ = 213%

Exercise 10.1

Write the following as percentages.

1. $\frac{17}{100}$ = _____

2. $\frac{25}{100}$ = _____

3. $\frac{56}{100}$ = _____

4. $\frac{3}{100}$ = _____

5. $\frac{49}{100}$ = _____

6. $\frac{9}{100}$ = _____

7. $\frac{89}{100}$ = _____

8. $3\frac{12}{100}$ = _____

9. $7\frac{77}{100}$ = _____

10. $5\frac{3}{100}$ = _____

10.2 Percentage to fraction

Percent means "out of 100," so to change a percentage to a fraction:
1. Write the percent over 100.
2. Simplify.

> Example: Write 30% as a fraction
>
> $30\% = \dfrac{30}{100}$
>
> Simplify $\dfrac{30}{100} = \dfrac{3}{10}$

It is useful to know the following percentages:

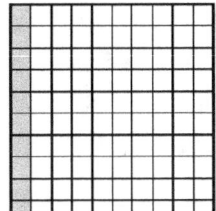 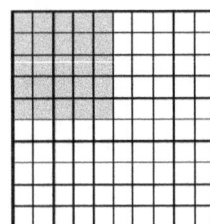 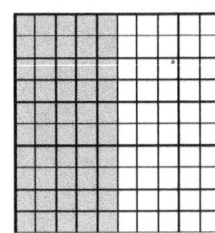 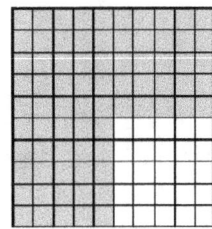

$10\% = \dfrac{1}{10}$ $25\% = \dfrac{1}{4}$ $50\% = \dfrac{1}{2}$ $75\% = \dfrac{3}{4}$

If we know that $10\% = \dfrac{1}{10}$, then we can use this to work out multiples of 10%.

> Example: Write 70% as a fraction.
>
> $70\% = 7 \times 10\%$
>
> $10\% = \dfrac{1}{10}$
>
> So $70\% = \dfrac{7}{10}$

Exercise 10.2

Write the following percentages as fractions.

1. 37% =

2. 23% =

3. 60% =

4. 25% =

5. 50% =

6. 24% =

7. 35% =

8. 5% =

9. 580% =

10. 172% =

10.3 Fraction to percentage

To convert fractions to percentages.

1. Multiply by 100
2. Cancel.

Example: Convert $\frac{3}{10}$ to a percentage.

Multiply by 100 ($\frac{100}{1}$)

$$\frac{3}{10} \times \frac{100}{1}$$

Cancel and multiply across.

$$\frac{3}{\cancel{10}_{1}} \times \frac{\cancel{100}^{10}}{1} = 30\%$$

Exercise 10.3

Convert the following fractions to percentages.

(If the fraction is worth more than one, then the percentage will be more than 100%).

1. $\frac{7}{10}$ = _____

2. $\frac{3}{5}$ = _____

3. $\frac{6}{25}$ = _____

4. $\frac{11}{20}$ = _____

5. $\frac{13}{50}$ = _____

6. $\frac{19}{20}$ = _____

7. $\frac{23}{25}$ = _____

8. $\frac{8}{5}$ = _____

9. $1\frac{2}{5}$ = _____

10. $2\frac{1}{20}$ = _____

10.4 Decimals and percentages

To convert decimals to percent multiply by 100.

To multiply a decimal by 100 move the decimal point two places to the right.

Example: Change 0.13 into a percentage.

0.13 x 100 = 13

Answer: 13%

To convert percentages to decimals divide by 100.

To divide a number by 100, move the decimal point two places to the left.

> Example: Change 13% into a decimal.
>
> 13 ÷ 100 = 0.13

This is exactly the same as fractions.

- To convert a percentage to a number (decimal or fraction), divide by 100.
- To convert a number (decimal or fraction) to a percentage, multiply by 100.

Exercise 10.4

Convert the following decimals to percentages.

1. 0.73 = _____

2. 0.25 = _____

3. 0.47 = _____

4. 0.03 = _____

5. 0.99 = _____

Convert the following percentages to decimals.

6. 12% = _____

7. 74% = _____

8. 27% = _____

9. 50% = _____

10. 6% = _____

10.5 Percentage of an amount

To calculate the percentage of an amount, work out the fraction and then multiply by 100.

> **Example:**
>
> 10 out of 25 biscuits have been eaten.
> What percentage have been eaten?
>
> First work out the fraction.
>
> $$\frac{10}{25}$$
>
> Without simplifying first, multiply by 100.
>
> $$\frac{10}{25} \times \frac{100}{1}$$
>
> Cancel and multiply across.
>
> $$\frac{10}{\cancel{25}_1} \times \frac{\cancel{100}^4}{1} = 40\%$$

Exercise 10.5

1. What percentage is 12 out of 48? _____

2. A cake is cut into 45 pieces. If 9 pieces of a cake have been eaten. What percentage have been eaten? _____

3. Julia has read 81 pages out of a book with 180 pages. What percentage of the book has she read? _____

4. Nethmi's shop came to $75. It was reduced by $12. What percentage reduction did the store give her? _____

5. A 500g block of cheese contains 150g of fat. What percentage fat is the cheese? _____

6. Ignacy got 24 out of 25 in a recent science test. What was his percentage? _____

7. In a recent maths test, 16 questions out of 40 were multiplication. What percentage of the test was multiplication? _____

8. A survey showed that 35 out of 84 people prefer salt and vinegar crisps. What percentage preferred salt and vinegar crisps? _____

9. Gill scores 48 out of 64 in a maths test. What percentage did she score? _____

10. At a shop there were 40 chocolate bars, but only 2 strawberry flavoured ones. What percentage of chocolate bars in the shop are strawberry flavoured? _____

Chapter 11 Calculating with Percentages

11.1 Calculating the amount

To calculate the amount:

- Write the percent over 100. (Percent means out of 100).
- Change the word "of" to times.
- Multiply.

Example: What is 48% of 25?

Percent means out of 100

$\frac{48}{100}$ of 25

Change 'of' to 'x'

$\frac{48}{100} \times \frac{25}{1}$

Cross cancel and multiply across

$\frac{\cancel{48}^{12}}{\cancel{100}_{4}} \times \frac{\cancel{25}^{1}}{1} = 12$

Exercise 11.1

Calculate the following:

1. 40% of 20 = _____

2. 20% of 50 = _____

3. 90% of 500 = _____

4. 30% of 24 = _____

5. A chocolate bar is 30% fat. How much fat is there in a 150g block?
   _____

6. A person buys a piece of furniture for $450, but needs to pay an additional 20% of the price as VAT. How much VAT do they need to pay?
   _____

7. Elsie is buying a hoodie, which costs $32, but there is a discount of 25%. How much is the discount?
   _____

8. There is a walking track 200km long. If Dante manages to walk 75% of the track, how far has he walked?
   _____

9. There are 180 stepping stones in a forest. If 85% of them are still above water how many are below water?
   _____

10. The Jones family have 250 music CDs. If 36% of them are classical music, how many of the CDs are classical music?
    _____

11.2 Increasing and decreasing by a percentage

Amounts can be increased by a percentage.

There are two methods:

First method
1. Calculate the amount of the percentage increase.
2. Add on to the original amount.

Second method
1. Add the percentage on to 100.
2. Calculate the percentage.

Example: A shop owner buys a computer for $200 and adds 30% profit. How much does he sell the computer for?

Method 1

Calculate 30% of $200

$$\frac{30}{100} \times \frac{200}{1}$$

Cancel and multiply

$$\frac{30}{\cancel{100}} \times \frac{\cancel{200}}{1} = 60$$

Add on to original amount

$200 + 60 = 260$

He sells the computer for $260.

Method 2

Add 30% on at the beginning

$100\% + 30\% = 130\%$

Calculate 130% of 200

$$\frac{130}{100} \times \frac{200}{1}$$

Cancel and multiply

$$\frac{130}{\cancel{100}} \times \frac{\cancel{200}}{1} = 260$$

He sells the computer for $260.

Amounts can also be decreased by a percentage.

There are two methods:

First method
1. Calculate the amount of the percentage decrease.
2. Subtract from the original amount.

Second method
1. Take away the percentage from 100.
2. Calculate the percentage.

Example: Annalise bought a pair of shoes for $40. She has a 15% off voucher. How much does she need to pay?

Method 1

Calculate 15% of 40

$$\frac{15}{100} \times \frac{40}{1}$$

Cancel and multiply

$$\frac{\cancel{15}^{3}}{\cancel{100}_{5}} \times \frac{\cancel{40}^{2}}{1} = 6$$

Subtract from original amount

$40 - 6 = 34$

She needs to pay $34

Method 2

Subtract 15% from 100

$100\% - 15\% = 85\%$

Calculate 85% of 40

$$\frac{85}{100} \times \frac{40}{1}$$

Cancel and multiply

$$\frac{\cancel{85}^{17}}{\cancel{100}_{5}} \times \frac{\cancel{40}^{2}}{1} = 34$$

She needs to pay $34

Exercise 11.2

1. Increase 180cm by 25% _____

2. Increase 460m by 30% _____

3. Decrease 80p by 45% _____

4. Decrease 440 km by 5% _____

5. A table has a length of 120cm but can be extended by 15%. What is the maximum length of the table? _____

6. A supermarket has a sign on its rice "an extra 20% free." If the original bags were 20kg, what size are they during the promotion? _____

7. A kitten in a pet shop cost $35. It was then reduced by 40%. What was the new price? _____

8. A DVD player cost $70. How much would the DVD player cost during the annual 20% off sale? _____

9. Some wood on a building site is damaged during a storm. There was 520 kg of wood and 30% was damaged. How much wood do they need to replace? _____

10. A shop buys badminton racquets for $20 and sells them for 45% profit. How much do they sell the racquets for? _____

11.3 Original Number

To calculate the original number:

1. Determine what percentage, the amount given in the question is.
2. Work out 1%
3. Multiply by 100

> Example: Eileen bought a skirt in a 30% off sale for $42. What was the original cost of the skirt?
>
> Work out 1% (do not simplify yet).
>
> If she bought it at 30% off, the skirt was 70% of the original price. If $42 is 70% then 1% will be
>
> $$\frac{42}{70}$$
>
> Multiply by 100
>
> $$\frac{42}{70} \times \frac{100}{1}$$
>
> Cancel and multiply across
>
> $$\frac{\cancel{42}^{6}}{\cancel{70}} \times \frac{\cancel{100}}{1} = \$60$$
>
> The skirt was originally $60.

Exercise 11.3

Work out the original amount for the following.

1. 60% is 360 cm _____
2. 90% is 270 g _____
3. 120% is $48 _____
4. 150% is 75 litres _____
5. 20% is 320 cm _____
6. 110% is 550 km _____

7. David bought a CD in a sale for $12. In the sale the price was reduced by 20%. What was the price of the CD before the sale? _____

8. The price of a computer including GST is $360. What is the cost of the computer for a charity which does not need to pay the 20% GST?

_____

9. The price of a bottle of perfume is reduced by 10%. If it now costs $27, what was the original price? _____

10. A house increases in value by 25% to $80 000. What was its value before the increase? _____

11.4 Percent increase and decrease

To work out the percentage an amount has been increased or decreased by:

1. Work out the increase or decrease – find the difference by subtraction.
2. Put this as a fraction over the *original* amount.
3. Multiply by 100.

Example: A $300 airfare is increased to $345. What is the percentage increase?

First work out the increase

345 − 300 = 45

Put over the original amount

$$\frac{45}{300}$$

Multiply by 100

$$\frac{45}{300} \times \frac{100}{1}$$

Cross cancel and multiply across.

$$\frac{\overset{15}{\cancel{45}}}{\underset{3}{\cancel{300}}} \times \frac{\cancel{100}}{1} = 15\%$$

Example: A $90 jacket is reduced to $72. What percentage has it been reduced by?

First work out the decrease

90 − 72 = 18

Put over original amount

$$\frac{18}{90}$$

Multiply by 100

$$\frac{18}{90} \times \frac{100}{1}$$

Cross cancel and multiply across.

$$\frac{\overset{2}{\cancel{18}}}{\underset{1}{\cancel{90}}} \times \frac{\cancel{100}}{1} = 20\%$$

Exercise 11.4

Work out the percentage increase or decrease for the following:

1. 800kg is decreased to 640 kg.   _____

2. 120g is decreased to 90g.   _____

3. 380 km is increased to 399 km.   _____

4. $90 000 was decreased to $72 000.   _____

5. 64 m is increased to 80 m.   _____

6. 80 km is reduced to 60 km.   _____

7. $32 is increased to $40.   _____

8. 800 kg is reduced to 120 kg.   _____

9. 65kg is decreased to 39 kg.   _____

10. 20 kg is increased to 42 kg.   _____

Chapter 12: Decimals

12.1 Decimal multiplication

To multiply decimals:
1. Remove the decimal points (completely ignore them).
2. Do the multiplication as normal.
3. Count the number of digits after the decimal point in the question.
4. Put the decimal point in so that the number of digits after the decimal point in the answer is the same as the number of digits after the decimal point in the question.

Example: Multiply 3.2 x 0.45

Remove the decimal points

32 x 45

Do the multiplication as normal

```
        3 2
    x   4 5
    ───────
        1 6 0
    1 2 8 0
    ───────
    1 4 4 0
```

Put the decimal point back in.

There are three digits after the decimal point in the question.

I need three digits after the decimal point in my answer.

The answer is: 1.430 or 1.43.

Exercise 12.1

1. 4.2 x 2.1 = _____

2. 2.6 x 3 = _____

3. 0.45 x 5 = _____

4. 0.4^2 = _____

5. 0.5 x 6 = _____

6. 8.2 x 0.4 = _____

7. 7.5 x 15 = _____

8. 0.5 x 0.03 = _____

9. 0.37 x 5.2 = _____

10. 0.004 x 87 = _____

12.2 Division of decimals

Unlike multiplication, we do **not** remove the decimal points when doing division.

In division the number we are dividing can contain decimal points without a problem. However, the divisor (the number we are dividing by) should not include decimals.

So that the divisor does not include decimals, multiply both sides of the divide by sign by the same amount. (Remember a divide by sign is the same as a fraction line).

6 ÷ 2 is the same as 60 ÷ 20 which is the same as 600 ÷ 200 .

Example: Calculate 89.25 ÷ 1.2

Multiply by 10 so that the divisor does not have a decimal point. (Do *not* multiply by 100 so that both numbers do not have a decimal point).

892.5 ÷ 12

Do division as normal

```
              7 4. 3 7 5
         1 2 | 8 9 2. 5 0
             - 8 4
               ─────
                 5 2
               - 4 8
                 ───
                   4 5
                 - 3 6
                   ───
                     9 0
                   - 8 4
                     ───
                       6 0
                     - 6 0
                       ───
                         0
```

89.25 ÷ 1.2 = 74.375

Exercise 12.2

1. 5.6 ÷ 0.7 = _____

2. 12 ÷ 0.03 = _____

3. 5.7 ÷ 1.2 = _____

4. 9.62 ÷ 0.8 = _____

5. $5.13 \div 0.6 =$ _____

8. $6.7 \div 0.25 =$ _____

6. $9.15 \div 1.5 =$ _____

9. $728 \div 1.3 =$ _____

7. $7.53 \div 0.12 =$ _____

10. $14.43 \div 0.1 =$ _____

12.3 Decimals to fractions

To convert decimals to fractions:

- Place the digits over the smallest place value
- Simplify

If converting a number greater than one, the units stay the same and just the decimal part is converted.

> Example: What is 1.375 as a fraction?
>
> The units become the whole number part of my fraction.
>
> The smallest place is the thousandths place.
>
> So put the digits after the decimal place over 1000.
>
> $$\frac{375}{1000}$$
>
> Simplify. It is easiest to simplify in two steps: dividing by 25 and then 5; or in three steps, dividing by 5 each time.
>
> $$\frac{375}{1000} = \frac{3}{8}$$
>
> Add to the units
>
> $$1.375 = 1\frac{3}{8}$$

Exercise 12.3

Convert the following decimals to fractions.

1. 0.7 = _____
2. 0.75 = _____
3. 0.48 = _____
4. 0.125 = _____
5. 0.85 = _____
6. 0.64 = _____
7. 1.35 = _____
8. 3.25 = _____
9. 1.45 = _____
10. 0.425 = _____

12.4 Fractions to decimal

A fraction line is the same as a "divide by sign."

So to convert a fraction to a decimal, do the division.

Example: Convert $\frac{1}{8}$ to a decimal.

Do the division

1÷8

```
        0. 1 2 5
    8 | 1. 0 0 0
        8
        ‾‾‾
          2 0
          1 6
          ‾‾‾
            4 0
            4 0
            ‾‾‾
              0
```

So $\frac{1}{8} = 0.125$

Exercise 12.4

Convert the following fractions to decimals.

1. $\frac{3}{10}$ = _____

2. $\frac{3}{4}$ = _____

3. $\frac{2}{5}$ = _____

4. $\frac{5}{8}$ = _____

5. $\frac{3}{5}$ = _____

6. $\frac{13}{100}$ = _____

7. $\frac{17}{40}$ = _____

8. $\frac{19}{20}$ = _____

9. $\frac{8}{25}$ = _____

10. $\frac{3}{16}$ = _____

12.5 Decimals, Fractions and Percentages

Summary of converting fractions, decimals and percentages.

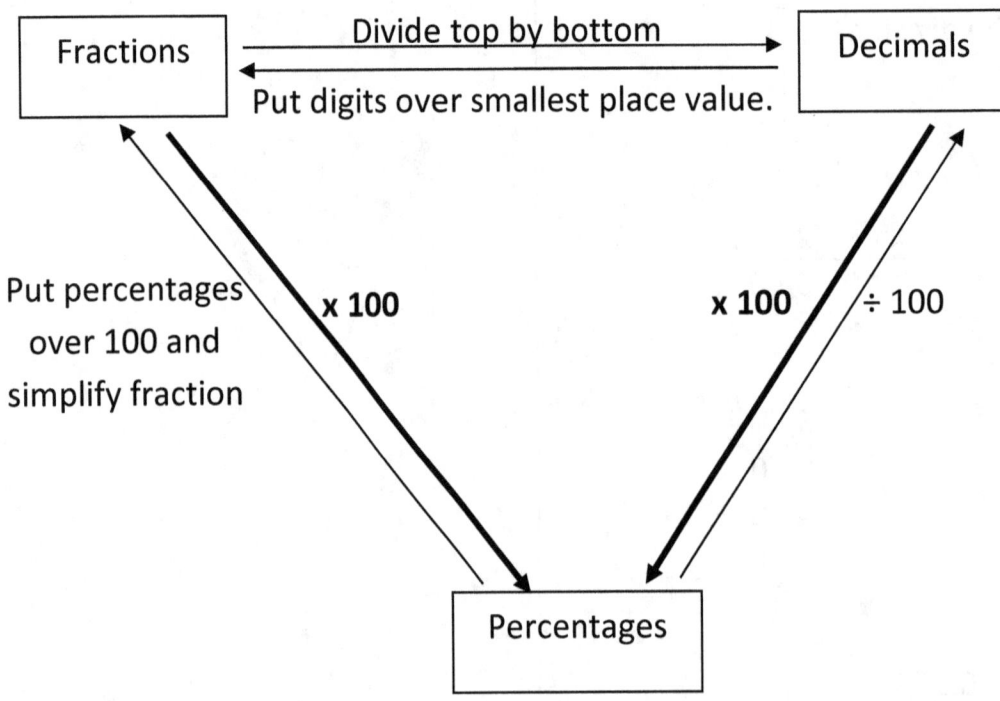

Examples:
1. Change 20% into a fraction.
   $20\% = \frac{20}{100} = \frac{1}{5}$

2. Change 0.13 into a percentage.
   0.13 x 100 = 13.   Ans: 13%

3. Change $\frac{3}{8}$ into a decimal.
   3 ÷ 8 = 0.375

4. Change 0.15 into a fraction.
   $0.15 = \frac{15}{100} = \frac{3}{20}$

Exercise 12.5

Complete the table below.

	Fraction	Decimal	Percentage
1.			42%
2.		0.365	
3.	$\dfrac{7}{20}$		
4.		0.24	
5.	$\dfrac{17}{25}$		
6.		0.63	
7.			65%
8.	$\dfrac{3}{20}$		
9.			31.5%
10.		0.1875	

Puzzle Page 2

Why did the student like maths?

$5782 + 364 + 97 - 12 =$ 6231   b

1. The equivalent of $\frac{14}{21}$ _____ __
2. Write $\frac{21}{8}$ as a mixed number   _____ __
3. $\frac{7}{16} \div \frac{1}{2} =$ _____ __
4. $\frac{1}{4} + 1\frac{1}{2} + \frac{1}{8} =$ _____ __
5. $4 - 1\frac{1}{8} =$ _____ __
6. Convert $\frac{75}{10}$ to a decimal   _____ __
7. Convert $\frac{3}{4}$ to a decimal   _____ __
8. Convert 0.57 to a percentage   _____ __
9. Convert $\frac{3}{15}$ to a percentage   _____ __
10. Convert 1.375 to a fraction   _____ __
11. $\frac{2}{5}$ of 60 =   _____ __
12. 45% of 60 =   _____ __
13. Percent increase from 90 to 130.5   _____ __
14. Percent decrease from 300 to 195   _____ __

Work out what letter each of your answers represents below:

24	0.75	7.5	27	20%	35%	57%	45%
n	w	o	l	c	p	t	f

6231	$\frac{2}{3}$	$1\frac{7}{8}$	$2\frac{7}{8}$	$\frac{7}{8}$	$1\frac{3}{8}$	$2\frac{5}{8}$
b	a	s	i	h	r	e

Now write the letter above the question numbers, below to answer the puzzle.

__ __   __ __ __   __ __ __ __ __ __
3   2   7   1   4   14   1   10   8   5   1   12

__ __   __ __ __ __ __ __ __ __
8   6   13   10   1   9   8   5   6   11   4

Chapter 13: Looking at data

13.1 Mean

To calculate the mean or average:

$$mean = \frac{total}{number}$$

That is to calculate the mean:
- Add up all the numbers
- Divide by how many numbers there are.

Example: What is the mean of $3\frac{1}{2}, 2\frac{1}{4}, 4\frac{1}{2}, 1\frac{3}{4}$

First, add up all the number: $3\frac{1}{2} + 2\frac{1}{4} + 4\frac{1}{2} + 1\frac{3}{4} = 12$

Second, we divide by 4 because there are four numbers.

$12 \div 4 = 3$. So the average of $3\frac{1}{2}, 2\frac{1}{4}, 4\frac{1}{2}, 1\frac{3}{4}$ is 3.

Exercise 13.1

Calculate the mean of the following numbers.
1. 5, 2, 8
2. 12, 15, 8, 5
3. 15, 20, 37, 28
4. 32, 23, 35
5. 4, 6, 8, 3, 5, 4
6. 4.5, 3.2, 1.3
7. 10.2, 12.7, 10.4
8. $5\frac{3}{8}, 2\frac{1}{2}, 4\frac{1}{8}$
9. $\frac{3}{5}, \frac{3}{10}, \frac{7}{10}, \frac{1}{2}, \frac{2}{5}$
10. 1127, 1389, 3476, 6234, 3210, 2564

13.2 Mean problems

If given the mean, we can calculate either the number of items or the total.

Total = mean x how many

Example: If the average cost of three items is $6, what is the total cost?

Total = mean x how many
= 6 x 3
= $18

Therefore, the total cost is $18.

An amount can also be found.
To do this, find the total and then subtract the other amounts.

> Example: Five girls go on a shopping trip. They spent an average of $15. If four of the girls spent $8, $20, $22 and $12, how much does the fifth girl spend?
> Add up the amounts spent by of the four girls: 8 + 20 + 22 + 12 = 62
> Work out the total: total = mean x how many
> = 15 x 5 = 75
> Subtract: Total – spend of other girls = 75 – 62 = 13
> The fifth girl spent $13.

The amount can also be worked out if the mean is given and the mean of one less is given. In this case work out both totals and subtract.

> Example: The mean score of 4 children on a test is 8. When Adi's score is added the mean is 7. What is Adi's score?
>
> Total of four = mean x how many = 8x4 = 32
>
> Total of five = mean x how many = 7 x 5 = 35
>
> Then subtract: 35-32 = 3
>
> Adi's score on the test was 3.

Exercise 13.2

1. The average cost of 6 books is $5.50. What is the total cost? _____

2. Another book is added to the books in question 1. This book cost $9. What is the mean cost of all 7 books? _____

3. The mean cost of 9 items is $8. What is the total cost? _____

4.

Colour	Number
Red	4
Blue	12
Purple	?
Green	5
White	5
Pink	2
Mean	6

Year 6 did a survey of their favourite colour. How many children liked purple best? The mean was 6.

_____

5. Mr Wong buys 7 items for $84. He then buys another item. The mean for all eight items is $11. How much was the final item he added? _____

6. The mean temperature for a week was 12°C. The temperatures are shown below:

Mon	Tue	Wed	Thu	Fri	Sat	Sun
10°C	7°C	9°C	13°C	14°C	?	17°C

What was the temperature on Saturday? _____

7. A group of friends each have an average of 45 stickers. A sixth person adds their stickers making the average 47. How many stickers do the children have altogether? _____

8. Shona attained the following scores (recorded in percent) in her maths tests: 73, 94, 86, 99. What is the lowest percentage she needs on her next maths test to have an average of 90%? _____

9. Frank wants to score 90% in Science. His Science mark is an average of Biology, Chemistry and Physics. If he scored 86% in Biology and 89% in Chemistry, how much does he need to score in Physics? _____

10. There are fifteen students in a dance performance. The mean age of 10 students is 16 years and the mean age of the other 5 students is 25 years. What is the mean age of all the dancers in the performance? _____

13.3 Mode and Median

There are three types of average:

- Mean – total divided by how many
- Mode – the most common number
- Median – the middle number

The mode is the most common number. This means that there can be more than one mode if two or more numbers appear the same number of times (but more than any other number). If all numbers appear the same number of times then there is no mode.

```
Examples:  3, 7, 7, 10, 12           mode is 7
           4, 4, 5, 12, 8, 11, 8     mode is 4 and 8
           4, 5, 7, 8, 11, 12        mode: none
```

To calculate the median, the numbers need to be placed in order from smallest to largest. The median is then the middle number. If there is an even number, then the median is the mean of the middle two numbers.

```
Examples:         13, 5, 7, 3, 2
Place in order:   2, 3, 5, 7, 13
Middle number: 5;  so median is 5


                  1, 13, 5, 7, 3, 2
Place in order:   1, 2, 3, 5, 7, 13
Middle numbers: 3 and 5
```
Mean of middle numbers: 4; so median is 4

Exercise 13.3

Calculate the mean, mode and median of the numbers below:

1. 4, 5, 8, 9, 4   mean: __   mode: ____   median: __

2. 12, 15, 11, 12, 20   mean: __   mode: ____   median: __

3. 9, 15, 9, 12, 11, 10   mean: __   mode: ____   median: __

4. 22, 24, 38, 46, 54, 38, 56, 42   mean: __   mode: __   median: __

5. 10, 13, 8, 13, 18, 10   mean: __   mode: ____   median: __

6. 9, 9, 9   mean: __   mode: ____   median: __

7. $2\frac{1}{2}, 3\frac{1}{4}, 5\frac{1}{2}, 2\frac{3}{4}$ mean: __   mode: ____   median: __

8. 2cm, 10cm, 6cm, 6cm   mean: __   mode: ____   median: __

9. 1.7m, 1.8m, 1.2m, 1.5m, 1.8m   mean: __   mode: __   median: __

10. 5.2kg, 8.3kg, 6.8kg, 5.1kg   mean: __   mode: ____   median: __

13.4 Range

The range is one calculation used to show the spread of the numbers. That is how far apart the numbers are.

While 3, 6, 1, 20   and   9, 10, 10, 11

both have a mean of 10, the range of the first set is 19 and the second set is 2.

The range is calculated using the equation:

Range = biggest number − smallest number

Example: Find the range of 22, 3, 7, 10, 8

The biggest number is 22 and the smallest 3;

so Range = 22-3 = 19

Exercise 13.4

Calculate the range of the following sets of numbers.

1. 12, 2, 5   Range: _____

2. 1, 3, 8, 2, 5   Range: _____

3. 27, 11, 4, 8   Range: _____

4. 12, 14, 15, 11   Range: _____

5. 208, 216, 197, 202   Range: _____

6. 0.1, 0.23, 0.04, 0.18   Range: _____

7. $5\frac{1}{2}, 1\frac{1}{4}, 9\frac{3}{4}, 3\frac{3}{4}$ Range: _____

8. 5.2km, 7.8km, 1.6km, 8.1km   Range: _____

9. 1.3cm, 1.7cm, 1.1cm, 0.8cm   Range: _____

10. 18.6kg, 27.4kg, 13.7kg, 24.3kg   Range: _____

Chapter 14: Money

14.1 Dollars and cents

There are 100 cents in one dollar.

$1 = 100c

Dollars are always written as a whole number or with two decimal places, the decimal place separating the dollars and the cents.

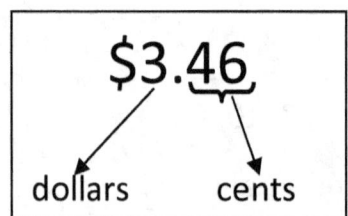

To convert dollars into cents, multiply by 100. Since to multiply by 100, the decimal point is moved two places to the right, this is the same as removing the decimal point.

To convert cents into dollars, divide by 100.

Exercise 14.1

Convert into cents.

1. $3.46   _____
2. $5.10   _____
3. $0.42   _____
4. $17.88 _____
5. $5   _____

Convert into dollars.

6. 600¢   _____
7. 752¢   _____
8. 380¢   _____
9. 203¢   _____
10. 32¢   _____

14.2 Calculating with money

Calculating with money is the same as any other calculation except:

- Need to convert to either all dollars or all cents first.
- All decimals must have two decimal places.
- Fractions are not used.
- If the answer is over 100¢, write as dollars.

Example: What is 10% of $15.

10% is the same as $\frac{1}{10}$

So, 10% of 15 is 1.5

However, need to use 2 decimal places

so, 10% of $15 is $1.50.

Example: Add $3.42, $12.76, 89¢, $1.12

First need to convert 89¢ to $0.89.

Then use column addition as normal.

```
         3 . 4 2
       1 2 . 7 6
         0 . 8 9
         1 . 1 2
    ───────────────
    $  1 8 . 1 9
```

Exercise 14.2

1. $12.23 − 76¢ = _____

2. $4.50 × 18 = _____

3. $500 ÷ 4 = _____

4. $3.24 + $7.82 + 21¢ = _____

5. 3 friends go together to buy a birthday cake for a friend. If the cake costs $10.20, how much does each friend need to contribute? _____

6. Ifeanyi is saving up for a computer game that costs $27.50. So far she has saved $12.80. How much more does she need to save? _____

7. For a party Alira buys 15 balloons at 30¢ each. How much does she spend on balloons? _____

8. Eric is saving up to buy a car that costs $2300. He saves $300 / month, from his part time job. How long will it take him to save for the car? _____

9. Stamps cost 53¢ each. Amber is writing her Christmas letter. If she sends it to all 25 people on her list, how much will the postage cost her? _____

10. Mr Sudar's shopping comes to $80.34 but he then adds another item costing $5.81. What will he need to pay now? _____

14.3 Costs

Often it is useful to compare the cost of an item in different quantities, to find out which one is the better value. The unit cost is the cost of each: ounce, pound, millilitre or gram. Sometimes cost per 100g or 100ml is used.

Often it is easiest to convert larger units to smaller ones and pounds to pence first.

Unit cost can be calculated using the formula:

Unit cost = cost of item ÷ quantity

Examples: 50g of coffee cost $1.50

$1.50 = 150¢

150¢ ÷ 50 = 3¢ per gram

6 bread rolls cost 90¢

90¢ ÷ 6 = 15¢ per roll

Exercise 14.3

Calculate the unit costs of the following and decide which one is the best value.

1. 500g sugar for 86¢; $1.42 for 1kg

   _____ / 500g   _____ / 500g   best value: _____

2. 100g block of chocolate for 87¢ or 300g block for $2.70

   _____ / 100g   _____ / 100g   best value: _____

3. 12 stickers for $1.20 or 150 stickers for $9.00

   _____ / sticker   _____ / sticker   best value: _____

4. Pack of 3 bars for 90¢ or pack of 8 bars for $2.16.
   _____ / bar   _____ / bar   best value: _____

5. 700g of rice for 91¢ or 10kg for $12
   _____ / 100g   _____ / 100g   best value: _____

6. A supermarket sells baked beans in two sizes. A 400g can costs 32¢ while a 700g costs 56¢.
   _____ / 100g   _____ / 100g   best value: _____

Star cereal costs

300g	500g	700g	1.5kg
63¢	$1.10	$1.40	$2.75

7. What is the unit cost for each of the Star Cereal boxes (to the nearest cent)?
   _____ / 100g   _____ /100g   _____ / 100g   _____ / 100g

8. Put them in order from best value to least value.
   _____   _____   _____   _____

9. What would be the lowest cost for buying exactly 10kg?   _____

10. What is the maximum amount of cereal I could buy for $20?   _____

14.4 Currency conversions

Conversions between currencies are done by:

1. Finding the exchange rate
   This is done by dividing
   If £5 = $4, then £1 = $\frac{4}{5}$ = $0.80   or $1 = $\frac{5}{4}$ = £1.25
2. Then multiply to find the value.
   Example: If a mug of chocolate costs $2, what is the cost of the hot chocolate in pounds?
   $1 = £1.25,   So $2 = 2 x 1.25 = £2.50

Often it is easiest to convert dollars to cents or pounds to pence before dividing in step 1. For simplicity, two currencies that both use dollars are not used in the same question. Obviously this sometimes happens. In this case it is important to clearly distinguish between the two.

Exercise 14.4

1. Giacomo is visiting England and going on holiday to Italy. The exchange rate is £1 = €1.30. How many Euros will he get for £350?

2. On his return, he has €260 remaining. How many pounds is this worth? (The exchange rate is still £1 = €1.30).

3. Giacomo then decided to visit Switzerland and exchanges €80 into Swiss Francs. How many Swiss francs would he get? €1 = 1.20 Francs.

4. Peter and Yvonne went on holidays to the UK. They bought a T-shirt for £5 which was worth $12. What is the exchange rate? (answer in the form: £1 = $___) _____

5. They later went out for a meal, which cost $35. How much did their meal cost in pounds? _____

6. Sheila was buying gifts in Australia to send to England. She looked at a UK online store and found a camera costing £70. How many Australian dollars would it cost her? (£1 = $1.80) _____

7. She bought a hamper costing $135. How much was the hamper in pounds? _____

8. Yinling is going to Taiwan. The exchange rate is £1 = $45 How many pence is one dollar worth? (round to 1dp) _____

9. She converts £500 into Taiwan dollars. How many Taiwan dollars does she have? _____

10. She buys a bowl for $60. How many pounds does the bowl cost? _____

Chapter 15: Time

15.1 As time goes by

Some units of time you need to know:

60 seconds = 1 minute	365 days = 1 non-leap year
60 minutes = 1 hour	366 days = 1 leap year
24 hours = 1 day	52 weeks (and 1 day) = 1 year
7 days = 1 week	10 years = 1 decade
2 weeks = 1 fortnight	100 years = 1 century
12 months = 1 year	1000 years = 1 millennium

The four seasons are: Summer, Autumn, Winter, Spring.

Exercise 15.1

1. How many days in a fortnight? _____

2. How many seconds in 3 minutes? _____

3. How many hours in $\frac{1}{3}$ of a day? _____

4. How many hours in a week? _____

5. How many minutes in 5 hours? _____

6. How many centuries in 2 millenia? _____

7. How many weeks in a decade? _____

8. How many minutes in 2 days? _____

9. How many hours in 510 minutes? _____

10. If all the seasons span the same length of time, how many weeks are there in Winter? _____

15.2 12 hour clock

On a clock the short hand represents the hours and the long hand represents the minutes. If there is a third thin hand, this represents the seconds.

There are four special positions on the clock. If the long hand is on the:

- 12 it is something o'clock
- 6 it is half-past
- 3 it is quarter past
- 9 it is quarter to

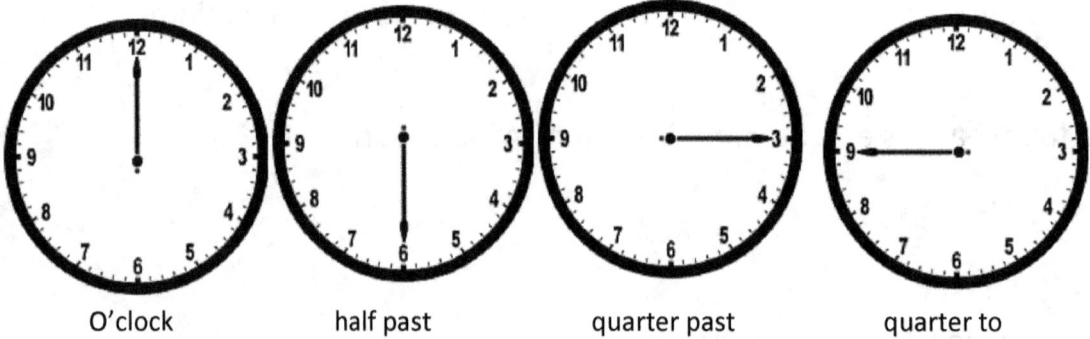

| O'clock | half past | quarter past | quarter to |

If the minute hand is after the 12 but before the six, we say it is past the hour, while if it is after the six but before the 12 we say it is to the next hour.

Every number on the clock represents 5 minutes.

Time can also be written in the same format as a digital clock. When written digitally, the hour is always the last hour that has passed and the minutes how many minutes since the hour.

Example: What is 13 minutes to seven am in digital form?

If it is 'to' seven, it is not yet seven, so the hours is 6. It's 13 minutes to the next hour, so the minutes are 60-13 = 47.
So the time is 6:47am

In 12 hour time, the time increases from midnight to 12:00 midday, then goes back to one and increases again to 12:00 midnight. From midnight to midday am is written after the time. From midday to midnight pm is written after the time.

- am stands for ante meridiem which means before midday.
- pm stands for post meridiem which means after midday.

It is useful to remember:

- 12 midday is 12:00 pm
- 12 midnight is 12:00 am

12 midday can also be written as noon, or 12 noon.

Example: This time was shown on the clock on Tuesday morning. Write the time in standard and digital format.
Standard: Thirteen minutes past eight am
Digital: 8:13am

Exercise 15.2

Write the following times in standard and digital format.

1. Morning

2. Evening

3. Time for lunch

4. Afternoon

Standard: _____
_____
_____

Digital: _____

Standard: _____
_____
_____

Digital: _____

Standard: _____
_____
_____

Digital: _____

Standard: _____
_____
_____

Digital: _____

5. Afternoon

6. Afternoon

7. Early morning

8. Nearly midday

Standard: _____
_____
_____

Digital: _____

Standard: _____
_____
_____

Digital: _____

Standard: _____
_____
_____

Digital: _____

Standard: _____
_____
_____

Digital: _____

9. Morning

10. Afternoon

Standard: _____
_____
_____

Digital: _____

Standard: _____
_____
_____

Digital: _____

15.3 24 Hour Time

In 24 hour time the hours continue after 12 midday, to 13. Twenty four hour time is always written with four digits and *no* am or pm.

To convert from 12 hour time to 24 hour time:

- If it is am, write as four digits and remove the am
  Example: 8:36am in 24 hour time is 08:36.
- If it is pm, add 12 to the hours and remove the pm.
  Example: 8:36pm in 24 hour time is 20:36.

To convert from 24 hour time to 12 hour time:

- If it is less than 12 (i.e. 11 or below in the hours) then remove any leading zero and add am
- If it is 12, then leave as is and add pm
- If it is more than 12, then subtract 12 from the hours and add pm

One final thing to remember: midnight

- In 12 hour time is 12:00am
- In 24 hour time is 00:00.

Exercise 15.3

Write the following in 24 hour time.

1. 7:22 am   _____
2. 11:56 am   _____
3. 5:24 pm   _____
4. 3 minutes after midnight.   _____
5. Quarter to 10 in the evening.   _____

Write the following 24 hour times in digital 12 hour time.

6. 12:06 _____

7. 18:30 _____

8. 06:50 _____

9. 00:20 _____

10. 21:15 _____

15.4 Time fractions

Before working out any fractions, percentages or time problems, all numbers must be in the same units.

Example: What fraction of a week is 42 hours.
First convert the week to hours. 1 week = 24 x 7 = 168 hours.
So the fraction is $\frac{42}{168}$. Simplify $\frac{42}{168} = \frac{1}{4}$.
So 42 hours is $\frac{1}{4}$ of a week.

Exercise 15.4

1. What fraction of a minute is 45 seconds? _____
2. What fraction of a minute is 10 seconds? _____
3. What fraction of an hour is 15 minutes? _____
4. What fraction of a day is 8 hours? _____
5. What fraction of a day is 15 hours? _____
6. What fraction of a week is 14 hours? _____
7. What percentage of a day is 12 hours? _____
8. What percentage of a day is 6 hours? _____
9. What percentage of a year is 3 months? _____
10. What percentage of a minute is 20 seconds?

    (round to the nearest percent). _____

Chapter 16: Calculating Time

16.1 Adding time

When adding or subtracting time, use normal column addition and subtraction but be aware of the difference in value between columns. Always add the smallest units first.

Example: If a movie that lasts for 1 hour 20 minutes starts at 7:50pm and there are 25 minutes of ads through the movie, what time will it finish?

```
7 : 5 0
1 : 2 0
    2 5
_____
```

Always start by adding the smallest place value, so add the minutes first.

```
7 : 5 0
1 : 2 0
    2 5
_____
      5
```

When the tens place of the minutes is added, 5+2+2=9. Six in this column equals one in the next column (because 60 minutes = 1 hour, so 6 in the tens place of the minutes is the same as one hour).
There is one lot of 6 in 9, so carry one to the next place value. There is 3 left.

```
  1
7 : 5 0
1 : 2 0
    2 5
_____
  : 3 5
```

Continue, by adding the hours.

```
    1
    7 : 5 0
    1 : 2 0
        2 5
    ─────────
    9 : 3 5
```

The movie would finish at 9:35pm

Example: Ayo is 6 years and 8 months old. How old will she be in 3 years and 6 months?

When adding remember 1 year = 12 months.

```
      6 y 8 m
　+   3 y 6 m
    ─────────
```

Start by adding months. 8+6=14. Every 12 months is 1 year. There is one lot of 12 in 14, so carry one. There is 2 remaining.

```
    1
      6 y 8 m
　+   3 y 6 m
    ─────────
            2 m
```

Then add the years.

```
    1
      6 y 8 m
　+   3 y 6 m
    ─────────
    1 0 y 2 m
```

So she will be 10 years and 2 months old.

Exercise 16.1

1. 4 hours 25 minutes + 8 hours 20 minutes _____

2. 3 hours 35 minutes + 2 hours 45 minutes _____

3. 6 hours 15 minutes + 4 hours 45 minutes _____

4. 3 days 5 hours + 7 days 19 hours _____

5. 11 days 11 hours + 2 days 16 hours _____

6. 9 years 7 months + 8 months _____

7. 12 years 8 months + 3 years 9 months _____

8. 3 weeks 6 days 15 hours +
   2 weeks 3 days 4 hours _____

9. 2 weeks 5 days 17 hours +
   1 week 1 day 9 hours _____

10. 7 hours 45 minutes 35 seconds +
    1 hour 20 minutes 45 seconds _____

16.2 Subtracting Time

When subtracting time do each unit in turn. Be careful what each unit is worth compared to adjacent units. Subtract each unit in turn, starting with the smallest unit. If there is larger number being subtracted borrow from the next unit before you subtract.

Example: 6 hours 10 minutes – 2 hours 30 minutes.

```
   6 : 1 0
 - 2 : 3 0
 ─────────
```

Start by subtracting the minutes.

30 is larger than 10, so borrow from the hours.

However, 1 hour = 60 min, so 1 in the hours place is worth 60 minutes.

10 + 60 is 70. So now subtract 30 from 70 (70-30=40)

```
     5    7 0
     6̶ : 1̶ 0̶
   - 2 : 3 0
   ──────────
              0
```

Now, continue as normal.

```
     5    7
     6̶ : 1̶ 0
   - 2 : 3 0
   ──────────
     3 : 4 0
```

So, the answer is 3 hours and 40 minutes.

Example: 12 days, 6 hours – 2 days 10 hours

```
    12  d   6  h
-    2  d  10  h
```

Start by subtracting the hours.

10 is larger than 6, so borrow from the days.

1 day is worth 24 hours, so add 24 onto the hours.

```
    11        +24
    1̶2̶  d   6  h
-    2  d  10  h
```

Which gives:

```
    11        30
    1̶2̶  d   6̶  h
-    2  d  10  h
```

Now subtract.

```
    11        30
    1̶2̶  d   6̶  h
-    2  d  10  h
                20  h
```

Continue with the days.

```
    11        30
    1̶2̶  d   6̶  h
-    2  d  10  h
     9  d  20  h
```

Exercise 16:2

1. 3 hour 50 minutes – 2 hours 25 minutes   _____

2. 3 hours 40 minutes – 1 hour 55 minutes   _____

3. 2 hours 19 minutes – 45 minutes   _____

4. 3 days 2 hours – 8 hours   _____

5. 4 days 12 hours – 2 days 15 hours   _____

6. 3 weeks 5 days – 1 week 6 days _____

7. 8 weeks 2 days – 3 weeks 5 days _____

8. 6 weeks 3 days 2 hours -
   3 weeks 3 days 11 hours _____

9. 2 hours 12 minutes 25 seconds -
   1 hour 36 minutes 35 seconds _____

10. 3 hours 18 minutes 22 seconds -
    1 hour 49 minutes 45 seconds _____

16.3 Multiplying time

Time can also be multiplied and divided in the normal way, provided the value of adjacent columns is taken into account.

Example: If Sopé works 8 hours 30 minutes each day for 5 days, how much does he work in the week?

```
  8 h  3  0 min
  X    5
  ─────────────
            0
```

5 x 3 = 15. Every 6 in this column is worth one hour. There are two lots of six in 15, so 2 is carried. There is 3 left.

```
  8 h  3  0 min
  X    5
  ─────────────
      ⑮
      ²3  0
```

Continue as normal. 5 x 8 = 40. Plus the 2 that was carried gives 42.

```
  8 h  3  0 min
  X    5
  ─────────────
  4 2 h  ²3  0
```

So, Sopé worked 42 hours and 30 minutes during the week.

Exercise 16.3

1. 3 hours x 8   _____

2. 2 hours 5 minutes x 3   _____

3. 4 hours 20 minutes x 4 _____

4. 9 hours 45 minutes x 5 _____

5. 1 year 4 months x 3 _____

6. 2 years 4 months x 5 _____

7. 3 years 2 months x 9 _____

8. 3 days 4 hours x 4 _____

9. 2 days 5 hours x 6 _____

10. 4 days 11 hours 20 minutes x 5 _____

16.4 Dividing Time

Similarly division can be used. When doing division, write the remainder under the next column and add. Then continue the division, diving the total.

Example: Oyinkin is 3 times older than her brother. If she is 13 years 6 months, how old is her brother?

```
        4
    ┌─────────────
  3 │ 1 3 y    6 months
```

1 year is 12 months, so the remainder of 1 year becomes 12 when added to the months. Write this number under the months and then add.

```
        4
    ┌─────────────────────
  3 │ 1 3 y  +12  6 months
              1   2 months
              ─────────────
              1   8 months
```

Continue, as normal.

```
        4 y        6 months
    ┌─────────────────────
  3 │ 1 3 y  +12   6 months
               1   2 months
              ─────────────
               1   8 months
```

Her brother is 4 years and 6 months old.

Exercise 16.4

1. 4 hours 20 minutes divided by 2   _____

2. 4 hours divided by 3   _____

3. 5 hours 20 minutes divided by 4   _____

4. 8 hours 24 minutes divided by 6   _____

5. 3 days 2 hours divided by 2   _____

6. 8 days 12 hours divided by 6   _____

7. 19 days 16 hours divided by 8   _____

8. 4 years 2 months divided by 5   _____

9. 12 years 1 month divided by 5   _____

10. 5 hours 5 minutes 16 seconds divided by 4   _____

16.5 Once upon a time

Time questions are most commonly found in the form of word problems. Have a go at the questions below.

Exercise 16.5

1. Jennifer's grandfather celebrated his 64th birthday in 2010. In what year was he born? _____

2. Sam's grandmother was born in 1942. In what year will she turn 90? _____

3. Joe's sister is 2 years and 8 months younger than Joe. If it is Joe's 12th birthday, how old is his sister? _____

4. Fred is 3 times older than his younger sister. If his sister is 3 years and 2 months old, how old is Fred? _____

5. Angela is 4 times older than her younger brother. If she is 11 years and 4 months old, how old is her younger brother? _____

6. Maya worked 7 hours and 20 minutes a day for 6 days. How many hours did she work altogether? _____

7. Three brothers have a total of 7 hours 15 minutes to use the computer on a Saturday. If they share the computer time equally, how long does each brother have on the computer? _____

8. Jacob practices the piano for 45 minutes a day Monday to Friday and 2 hours 30 minutes on each of Saturday and Sunday. How much practice does he do in the week? _____

9. Nicole swims 100m in 50 seconds while Terry swims 100m in 65 seconds. By what percentage is Terry slower than Nicole? (You are comparing to Nicole so use 50 as the denominator) _____

10. Jim's times to complete an obstacle course are: 3 minutes 20 seconds, 5 minutes, 55 seconds and 4 minutes 15 seconds. What is his average time? _____

Chapter 17: Time Problems

17.1 Centuries

Centuries are counted from the year 1AD. AD stands for "Anno Domini," which means "in the year of the Lord," in Latin. Sometimes instead of AD, CE is used. CE stands for "Common Era."

Before 1AD was 1BC. BC stands for "Before Christ." Sometimes BCE is used. BCE stands for "Before Common Era." So 5BC is five years before the birth of Christ while 5000 BC is 5000 years before the birth of Christ.

While scholars think that they got the date of Christ's birth wrong it is still used as a reference point in calculating years.

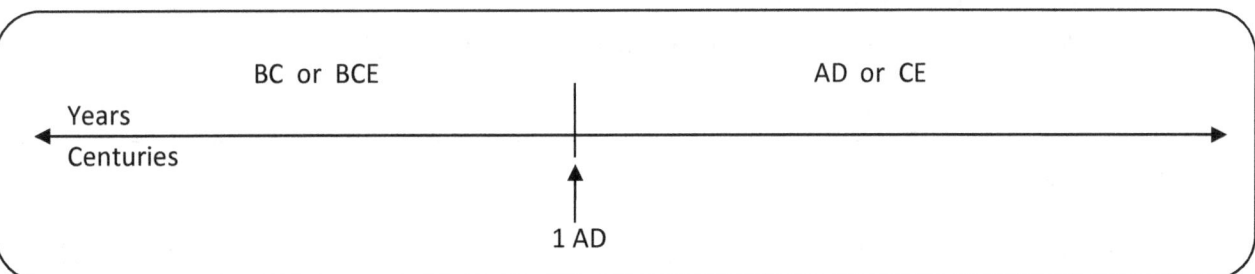

The first century AD was up to the year 99AD (technically 100AD, but these days the new century is seen as starting from the year ending with two zeros). So, the century a year is in, is the number of hundreds in the year plus one (we're adding on the first century before the date reached 100).

> Example: In what century was the year 1066?
> The number of hundreds is 10
> So the century is 10 + 1 = 11
> The year 1066 was in the 11th Century AD.

Exercise 17.1

Underline, which of the following occurred first (was longest ago).

1. 12BC, 13BC _____
2. 1BC, 5AD _____
3. 320BC, 35BC _____
4. 12AD, 13AD _____
5. 17th century AD, 1721 _____

Which century are the following years in?

6. 1684AD _____
7. 2006AD _____
8. 2017AD _____
9. 1823AD _____
10. 623BC _____

17.2 Logic time

In logic time questions always:

1. Start with the time given.
2. Work out each part in turn (do not combine parts).

Example:

Jemma's watch is 4 minutes slow. She has a 15 minute walk to school, which starts at 8:30am What is the latest time on her watch that she can leave for school?

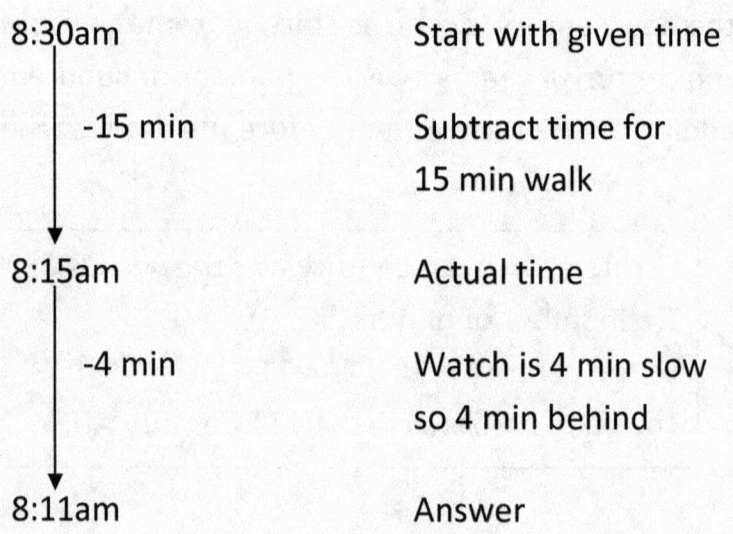

8:30am		Start with given time
-15 min		Subtract time for 15 min walk
8:15am		Actual time
-4 min		Watch is 4 min slow so 4 min behind
8:11am		Answer

Exercise 17.2

1. The 7:33am bus is 9 minutes late. If Harshavi's watch is 5 minutes fast, what time does the bus arrive by her watch? _____

2. If a coach trip from Town Centre to Town Edge takes 22 minutes but was delayed by roadworks for 6 minutes, what time did the bus arrive that left Town Centre at 0921? _____

3. A plane flight from Black Stump to Orange Beacon takes 50 minutes. However, tailwinds meant that the flight which left at 14:20 took 6 minutes less than normal. At what time did it arrive at Orange Beacon? _____

4. John's watch is 14 minutes slow. If his drive to work takes 35 minutes and he left at 7:20am at what time does he arrive, according to his watch? _____

5. The 3:40pm bus arrives 8 minutes early. If Priya's watch is 5 minutes fast, what time does Priya's watch say when the bus arrives? _____

6. Jeremy's watch is 15 minutes fast. According to his watch he leaves home at 8:30am What is the actual time he arrives if the journey takes 20 minutes? _____

7. Tighe takes 6 minutes to walk to school. He needs to be at school by 8:40am. What time must he leave home, by his watch, if his watch is 10 minutes fast? _____

8. Ela took an airplane from Heathrow to Amsterdam. The flight was scheduled to leave at 14:40 but was delayed by 40 minutes. If she was required to be at the airport two hours before departure, what time does she now need to be there? _____

9. Moses took the 1620 plane flight from Nairobi to Dodoma. He needed to arrive at the airport 90 minutes before the flight. If his journey to the airport takes 40 minutes, what time does he need to leave home? _____

10. The underground from Charing Cross to Baker Street takes 8 minutes. The service runs every 10 minutes. What is the latest that Stanley must get to the platform at Charing Cross to ensure he will reach Baker Street by 4:30pm? _____

17.3 Timetables

Make sure you understand what the question is asking. The question may require some addition or subtraction. Timetables are normally written using 24 hour time.

Exercise 17.3

Use the timetable below to answer the following questions:

Route 23									
Little Town Bus Station	07:27	07:41	07:51	08:02	08:11	08:21	08:36	08:41	09:06
Post office	07:36	07:50	08:00	08:11	08:20	08:30	08:45	08:50	09:15
St Barnabas	07:56	8:01	-	08:28	08:31	-	9:01	-	09:32
Ampfield Train Station	08:15	08:10	08:20	08:46	08:40	08:50	09:16	09:11	09:49
Tidmouth town centre	08:27	08:21	08:29	08:58	08:51	09:00	09:28	09:21	09:58

1. Which is the latest bus, from Little Town Bus Station, does Bob need to catch if he needs to be in Tidmouth before 9:00am? _____
2. How long does the bus take to get from Little Town Bus Station to the Post Office? _____
3. How long does the 07:41 bus from Little Town Bus Station take to get to Tidmouth? _____
4. Which bus takes the shortest amount of time to travel from Little Town Bus Station to Tidmouth? _____
5. Which bus takes the longest amount of time to travel from Little Town Bus Station to Tidmouth? _____
6. Celia needs to catch the 09:01 train from Ampfield Train Station. What is the latest bus she can catch from Little Town Bus Station? _____
7. If Bill needs to get to St Barnabas by 9:20am, what bus must he catch from Little Town Bus Station? _____

8. Ben has a 12 minute walk from the Tidmouth Town Centre stop to his work. If he needs to be at work by 8:40am, which bus does he need to catch from Little Town Bus Station? _____

9. Rihana catches the 08:02 bus from Little Town Bus Station, to the Post Office. She spends 10 minutes in the Post Office and then catches the next bus to Tidmouth Town Centre. At what time does she reach Tidmouth Town Centre? _____

10. Azzam needs to be at work at 9:00am for a meeting. He lives a 6 minute walk from the Post Office. His work is a 12 minute walk from the Ampfield Train Station. At what time must he leave home, to get to the meeting on time? _____

17.4 Time zones

Since the Earth is constantly rotating on its axis, it is different times in different parts of the world. So the world is divided into a number of time zones.

The standard time in England is referred to as Greenwich Mean Time (GMT). From this time, the time in other places can be calculated. Time zones can be shown on a map or in a table.

Timezones compared to Greenwich Mean Time (GMT)			
Timezone	Location	**Timezone**	Location
GMT +12	New Zealand	GMT + 0	UK, Morocco, Ghana
GMT + 11	Solomon Islands	GMT -1	Chad, Nigeria
GMT +10	Sydney	GMT -2	Mid-atlantic
GMT + 9.5	Darwin	GMT -3	Brazil
GMT +9	Japan, Korea	GMT -3.5	Newfoundland
GMT + 8	Perth, Hong Kong	GMT -4	Santiago
GMT + 7	Indonesia	GMT -5	Lima
GMT + 6	Dhaka	GMT -6	Mexico, Central US
GMT + 5.5	Kolkata	GMT -8	Pacific Time
GMT + 5	Islamabad	GMT -9	Alaska
GMT +4.5	Kabul	GMT -10	Hawaii
GMT + 4	Abu Dhabi	GMT -11	Samoa
GMT +3.5	Tehran	GMT -12	Marshall Islands
GMT + 3	Baghdad, Moscow		
GMT +2	Finland, Israel		
GMT +1	Austria, France		

This table does not take into account daylight savings. In some places the clocks go back an hour in Autumn and forward an hour in Spring.

> Example. When it is 0900 in England, what time is it in Sydney?
> As Sydney is GMT +10, add 10 hours to the time in London.
> 0900 + 10h = 1900 or 7:00pm
>
> Example. When it is 2:00pm in Abu Dhabi, what time is it in Kolkata?
> Abu Dhabi is GMT +4 and Calcutta is GMT +5.5, so Kolkata is 1.5 hours in front of Abu Dhabi.
> 2:00pm + 1:30 = 3:30pm It is 3:30pm in Kolkata.
>
> Example. When it is 6:00pm in Perth, Australia; what time is it in London?
> Easiest to convert to 24 hour time first. 6:00pm = 1800
> Perth is GMT +8 (the time is 8 hours in front of GMT)
> To go back to GMT need to take away the 8 hours.
> 1800 – 8h = 1000. It is 10:00am in London.

Exercise 17.4

Answer the following questions using the table on the previous page.

1. When it is 4:00pm in Wellington, New Zealand; what time is it in London? _____
2. When it is 1700 in London, what time is it in Japan? _____
3. Then it is 3:10pm in London, what time is it in Kabul? _____
4. When it is 3pm in England, what time is it in Alaska? _____
5. When it is 1:00pm in Brazil, what time is it in Lima? _____
6. When it is 4:00am in Finland, what time is it in New Zealand? _____
7. When it is 9:00pm in Samoa, what time is it in Brazil? _____
8. When it is 8:00pm in Samoa what time is it in Tehran? _____
9. When is 12pm in Baghdad, what time is it in Hawaii? _____
10. If it is 9:00am in Japan, what time is it in Lima? _____

Puzzle Page 3

Why are Saturday and Sunday the strongest days?

5782 + 364 + 97 − 12 =   6231   c

1. What is the mean of 15, 17, 22?   _____ ___

2. What is the median of 16, 21, 18, 4?   _____ ___

3. What is the range of 3, 12, 19, 18?   _____ ___

4. The exchange rate is £1:$1.75. How many dollars in £300?   _____ ___

5. How many days in a non-leap year?   _____ ___

6. How many hours in three days?   _____ ___

7. How many minutes in $\frac{1}{3}$ hour?   _____ ___

8. 8:30pm + 2 hours 50 minutes =   _____ ___

9. 12:00am − 3 hours 12 minutes =   _____ ___

10. 9:30am + 22 minutes − 54 minutes =   _____ ___

11. 11:24am − 2 hours 36 minutes =   _____ ___

12. What is 6 years 3 months divided by 5, in months?   _____ ___

Work out what letter each of your answers represents below:

15	16	17	18	20	72	365
d	o	e	t	y	h	l

6231	525	8:48am	11:20pm	8:48pm	8:58am
c	k	w	r	a	s

Now write the letter above the question numbers, below to answer the puzzle.

__ __ __   __ __ __   __ __ __ __ __ __
 9 5 5   1 6 2   3 1 6 2 8 10

__ __ __   __ __ __ __   __ __ __ __
 9 8 2   11 2 9 4   12 9 7 10

Chapter 18: Measurement

18.1 Metric Units

All measurements should have a number and a unit.

The basic metric units are:

- Length – metre (m)
- Mass – gram (g)
- Volume – litre (l)

To make a unit bigger or smaller a prefix can be added.

Prefix	Symbol	Meaning	Example
Mega	M	x 1 000 000	A megalitre (Ml) is a million litres.
kilo	k	x1000	A kilogram (kg) is a thousand grams.
Basic unit			
deci	d	$\frac{1}{10}$	A decilitre (dl) is a tenth of a litre. (That means there is 10 dl in every litre).
centi	c	$\frac{1}{100}$	A centimetre (cm) is a hundredth of a metre. (That means there are a 100 cm in every metre).
milli	m	$\frac{1}{1000}$	A milligram (mg) is a thousandth of a gram. (That means there are 1000 mg in every gram).

When using symbols it is important to be careful if the symbol is a lowercase letter or a capital letter; a mg is very different to a Mg!

The larger the unit, the less of them there will be.

Exercise 18.1

1. 1kg = _____ g
2. 1l = _____ ml
3. _____ cl = 1l
4. _____ ml = 1cl
5. _____ mg = 1g
6. 100g = _____ cg
7. 1kl = _____ l
8. _____ mm = 1m
9. 1ml = _____ cl
10. 1kg = _____ mg

18.2 Metric Conversions

To find the conversion factor:

1. Work out how many smaller units there are in a large unit.
2. Work out how to go from the unit you have to the unit being converted into.
3. Do the same thing with your number.

> Example: Convert 7.4 l to centilitres.
>
> 1. 1l = 100cl
> 2. To go from l to cl we x100
> 3. 7.4 x 100 = 740cl

Exercise 18.2

1. 23g = _____ kg
2. 403cm = _____ m
3. 5.47 l = _____ cl
4. 7.5cm = _____ mm
5. 9520ml = _____ l
6. 2.1kg = _____ g
7. 6.3 dl = _____ ml
8. 8.4 m = _____ mm
9. 50.2 mg = _____ g
10. 3 kl = _____ ml

18.3 Imperial Units

Sometimes imperial units are still used in the United Kingdom.

Here are some common imperial units:

Mass

1 pound (lb) = 16 ounces (oz)

1 stone = 14 pounds (lb)

1 ton = 160 stones (st)

Volume (capacity)

1 quart = 2 pints (pt)

1 gallon = 8 pints

Length

1 foot (ft) = 12 inches (in)

1 yard (yd) = 3 feet (ft)

1 mile (m) = 1760 yards (yd)

Quantities

1 dozen = 12

1 gross = 12 dozen (144)

1 score = 20

Baker's dozen = 13

To convert from one imperial unit to another:

1. Write what we know (the conversion).
2. Work out how to go from the known unit to the unit being converted to.
3. Do the same with the number that is being converted.

> Example: Convert 1.5 lb to ounces.
> 1. 1 lb = 16 oz
> 2. To go from lb to oz x16
> 3. 1.5 x 16 = 24
> So 1.5 lb = 24 oz

Exercise 18.3

1. 1ft = _____ in
2. 6pt = _____ quart
3. 30 in = _____ ft
4. 9 yd = _____ ft
5. 210 lb = _____ stone
6. 6 lb = _____ oz
7. 320 stone = _____ ton
8. 16 quart = _____ pints
9. 10 pints = _____ gallons
10. 3 score years and 10 = _____ years

18.4 Metric – Imperial Units

It is also possible to convert between imperial and metric units. The conversions below are inaccurate, so the larger the numbers being converted the more inaccurate the conversion.

Here are some approximate conversions:

Length

2.5cm ≈ 1 inch

30 cm ≈ 1 foot

8km = 5 miles

1km = $\frac{5}{8}$ mile

Mass

30g ≈ 1 oz

1 kg ≈ 2.2 lb

Volume

1 litre ≈ 2 pints

4.5 litres ≈ 1 gallon

Converting between metric and imperial follows the same steps as before:
1. Write down the conversion.
2. Work out how to go from the known unit to the unit being converted to.
3. Do the same with the number that is being converted.

A note on the ton(ne): The metric tonne is 1000kg, while the imperial ton is 160 stone. One metric tonne is approximately equal to one imperial ton.

Exercise 18.4

1. 12 in = _____ cm
2. 6 ft = _____ m
3. 6 pt = _____ l
4. 8 gallons = _____ l
5. 5 kg = _____ lb
6. 44 lb = _____ kg
7. 24 km = _____ miles
8. 120 g = _____ oz
9. 200 km = _____ miles
10. 1 metre = _____ ft

Chapter 19: Using Measurement

19.1 Using scales

All measurements should have a number and a unit.

To take measurements we normally need to read scales. Scales are lines with marks on them. Any marks between numbers stand for an equal amount.

For example, look at the scale below:

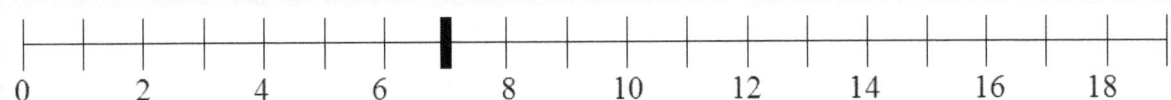

Every line is equal to 1. The reading on the scale is 7.

Now look at the scale below:

Every line is equal to 2, The reading on the scale is 26.

If the reading on a scale is not on a line then we need to estimate the distance between the lines and the mark.

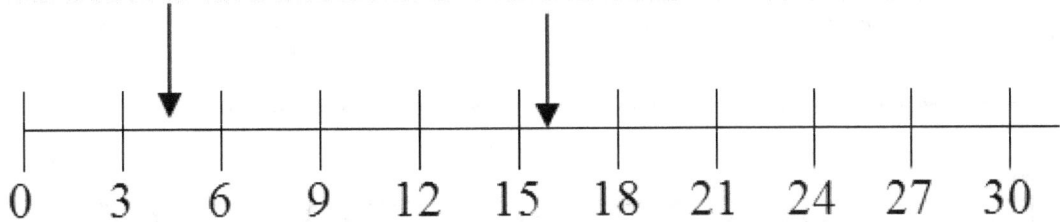

Every line is equal to 3. The first mark is half way between the 3 and the 6 so the reading on the scale is 4.5.

The second mark is one third the way between the 15 and the 18 so the reading on the scale is 16.

Example:

Read the following ruler in cm and in mm.

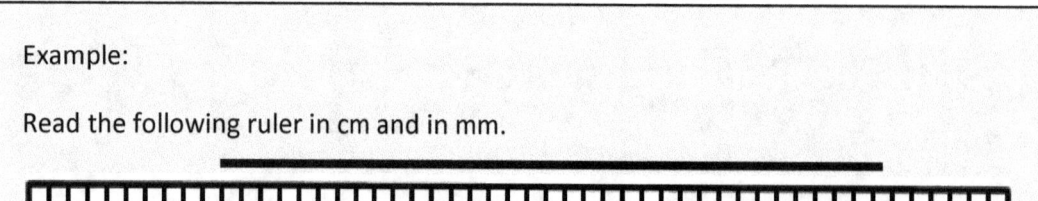

Centimetres are shown by the big lines, while millimetres (or 0.1cm) are shown by the small lines. The line goes from 1cm to after the 4, so there are 3 whole centimetres. There are then 4.5 millimetres. This gives 3.45cm. Each cm is worth 10mm, so this is equal to 34.5mm.
So my answer is:
3.45cm
34.5mm

Exercise 19.1

Read the following scales.

1.
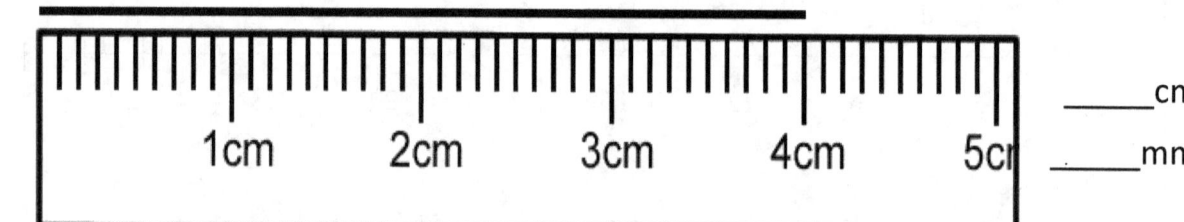
_____cm
_____mm

2.
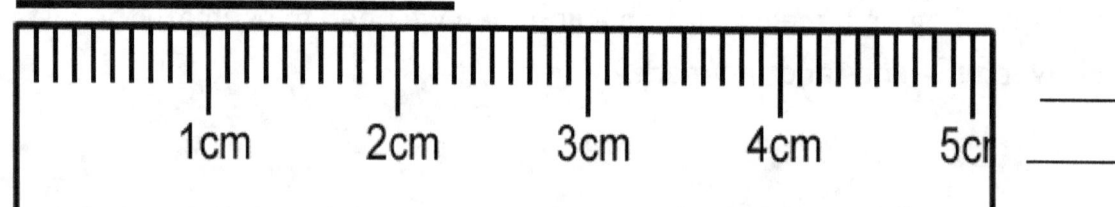
_____cm
_____mm

3.
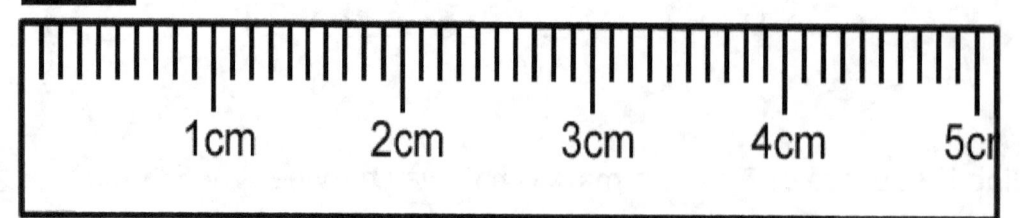
_____cm
_____mm

4. 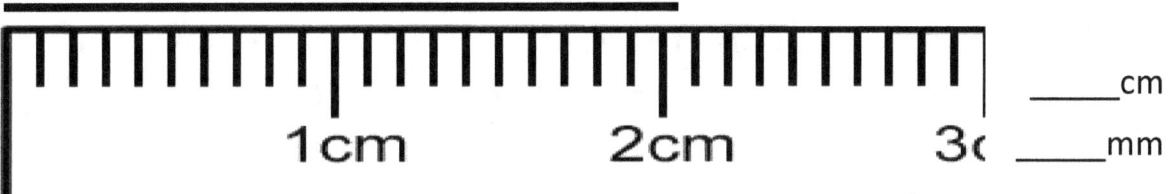 _____cm
_____mm

5. 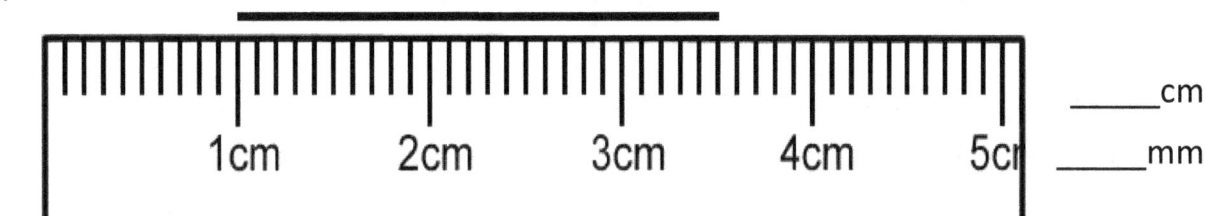 _____cm
_____mm

6.

_____ mph

7.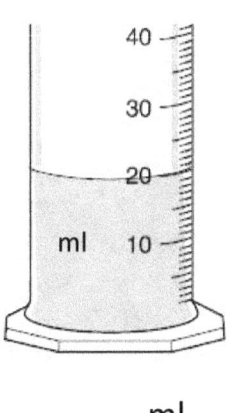

_____ ml

8.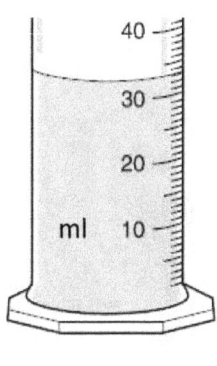

_____ ml

9.

_____ g
= _____ kg

10.

_____ g
= _____ kg

19.2 Temperature

Temperature is a measure of how hot or cold something is. Temperature is measured using a thermometer in degrees Celsius (°C).

While degrees Celsius (°C) is the standard metric unit, the imperial unit is degrees Fahrenheit. Sometimes degrees Kelvin is also used.

In degrees Celcius, the melting point of water is 0°C, the boiling point of water is 100°C and body temperature is 37°C.

Exercise 19.2

Read the following thermometers

1.

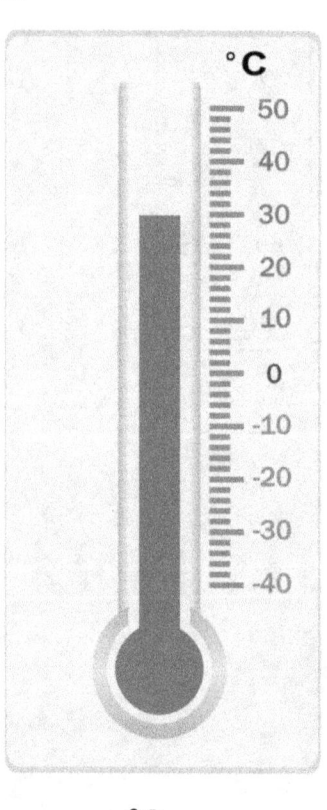

_____°C

2.

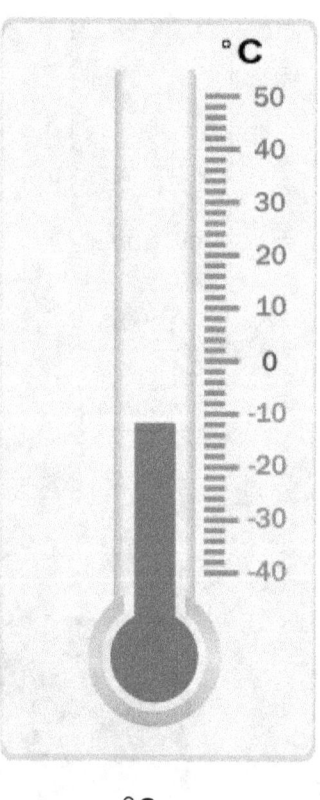

_____°C

3.

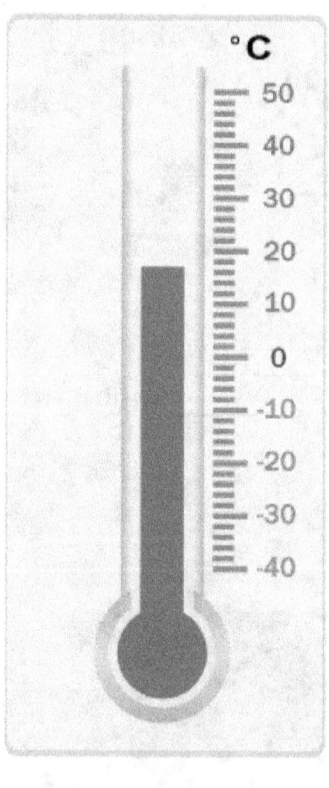

_____°C

4.

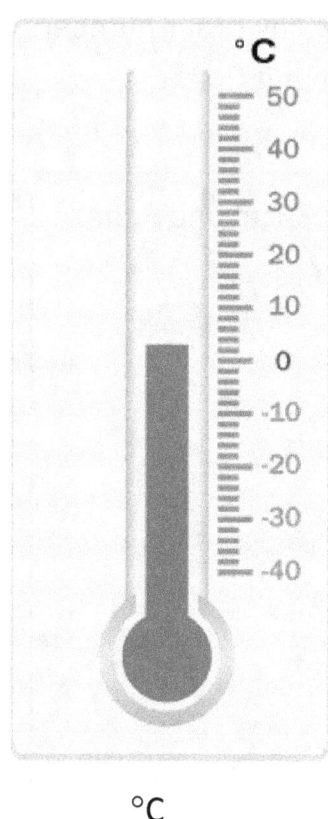

_____°C

5.

_____°C

Use the thermometers in questions 1-5 to answer the questions below:

6. Which thermometer shows body temperature? _____

7. Which thermometer is closest to the melting point of water? _____

8. If the thermometers were measuring water, for which thermometer is that water solid (ice)? _____

9. What is the difference between thermometers 2 and 3? _____

10. What is the difference between thermometers 3 and 4? _____

19.3 Calculating with measurements

Calculating with measurements is the same as calculating any other numbers, except all measurements must be converted to the same unit first.

> Example: Erin has three lengths of fencing: 5.4m, 80cm and 3.6m. What is the longest length of fence she can make?
>
> First need to convert 80cm to metres.
>
> 80cm = 0.8m
>
> Then add up the three lengths.
>
> ```
>     5 . 4
>     0 . 8
>   + 3 . 6
>   -------
>     9 . 8
> ```
>
> So the maximum length is 9.8m.

Exercise 19.3

1. Leo has a delivery to send that has 7 boxes which weigh 3.5kg each and 4 boxes which weigh 900g each. What is the total weight of the delivery?   _____

2. Frank has 2 jugs of squash. The large jug holds 1.7 litres and the small jug holds 600ml. If both jugs are full, how much squash is there?   _____

3. Rebecca is wrapping Christmas gifts. She has a 5 metre roll of ribbon. She uses 80cm on each of two gifts and 2.7m on a very large gift. How much ribbon does she have left?   _____

4. Niveka has a 2l bottle of soft drink. How many cups can she fill if each cup holds 150ml of drink? _____

5. Dean is going away on holiday. He is allowed 18kg of luggage. If his suitcase weighs 1.7kg, how much can he pack? _____

6. Liam is 30cm shorter than his mum. If his mum's height is 1.78m, what is Liam's height? _____

7. If apples have an average mass of 100g, how many apples would there be in a 10kg bag? _____

8. Deborah is making a large batch of shortbread. If the recipe uses 900g of flour, how much flour would she need for 7 batches? _____

9. John goes on a charity bike ride and rides 840km in 7 days. On average, how far does he ride each day? _____

10. If Sri has 3m of cellophane. How many gifts can she wrap, if she uses 15cm for each one? _____

19.4 Coordinates

Coordinates are values that show a position on a grid.

A grid normally has two axes. The horizontal axis is called the x-axis and the vertical axis is called the y-axis. These axes can be extended to include negative numbers. The point where the axes meet at (0,0) is called the origin.

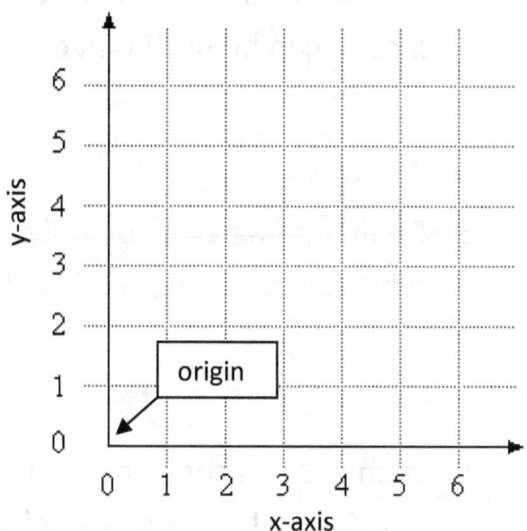

Coordinates are always written with the x-axis followed by the y-axis. There are a number of ways to remember that the x-axis is first:

- x comes before y in the alphabet
- you go down the corridor and up the stairs
- you crawl before you walk

Example: What are the coordinates of the point shown below?

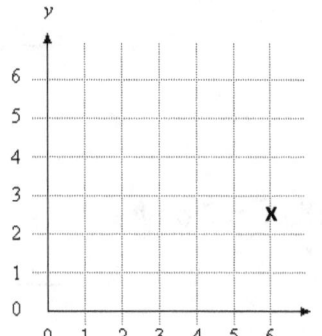

x-axis is first, the x-coordinate is 5

y-coodinate is 2.5

Therefore, the coordinates are (5, 2.5)

Exercise 19.4

Write the coordinates of the points below:

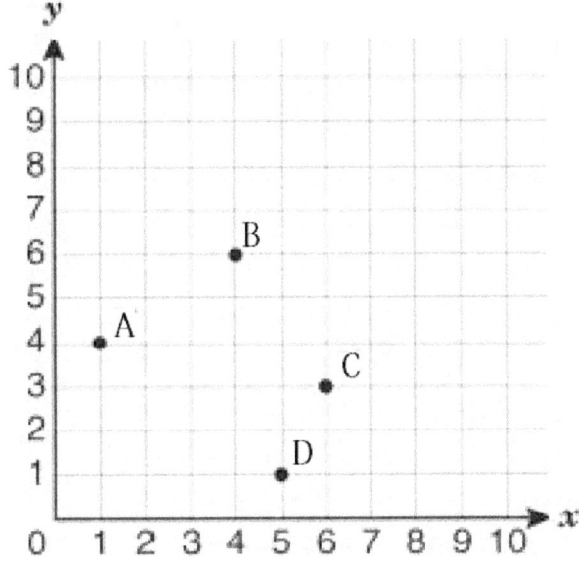

1. A _____
2. B _____
3. C _____
4. D _____

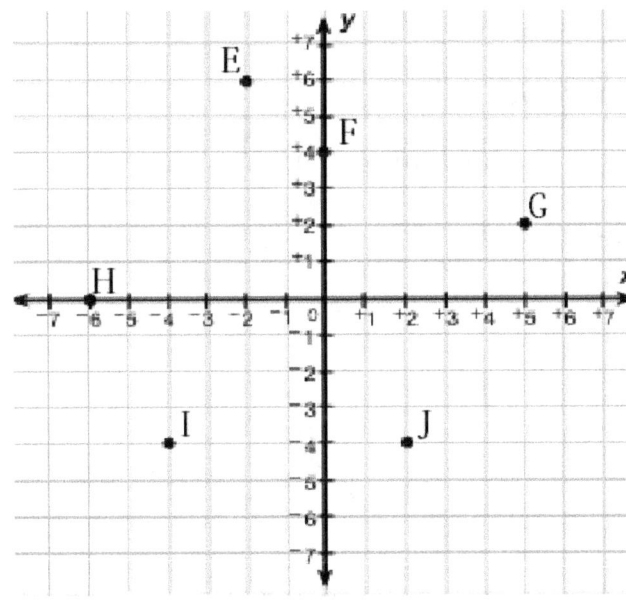

5. E _____
6. F _____
7. G _____
8. H _____
9. I _____
10. J _____

Chapter 20: Lines

20.1 Types of line

Lines can be horizontal, vertical or diagonal.

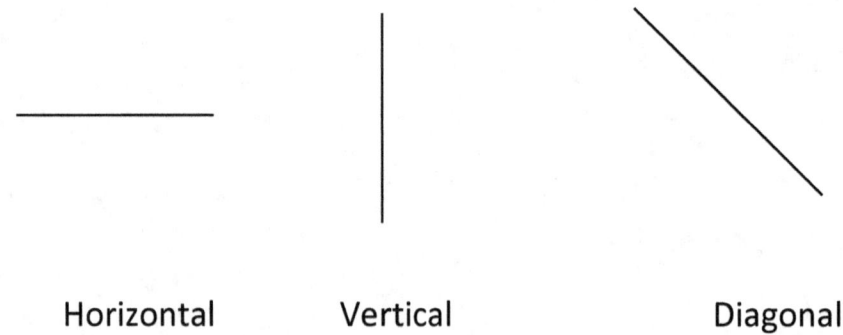

Horizontal   Vertical   Diagonal

Two lines can be related in a number of ways:
- Perpendicular – at right angles (do not have to meet).
- Parallel – remain the same distance apart.
- Neither.

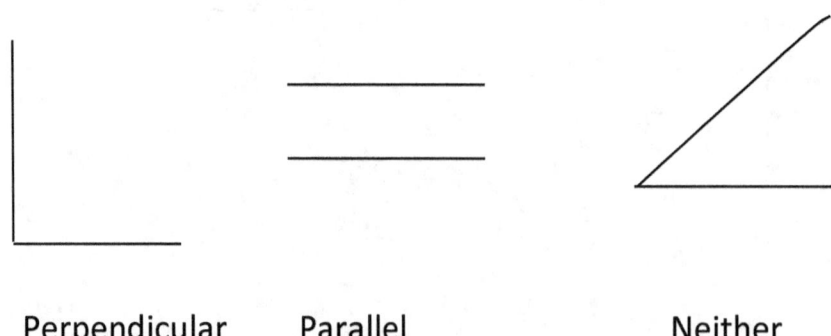

Perpendicular   Parallel   Neither

When two lines meet at a right angle (are perpendicular), this is depicted with a small square in the corner.

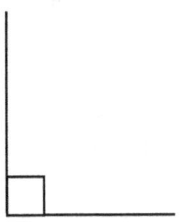

There are five different types of angles:

- Acute - less than 90°.
- Right - 90°.
- Obtuse – greater than 90° but less than 180°.
- Straight - 180°.
- Reflex – greater than 180°.

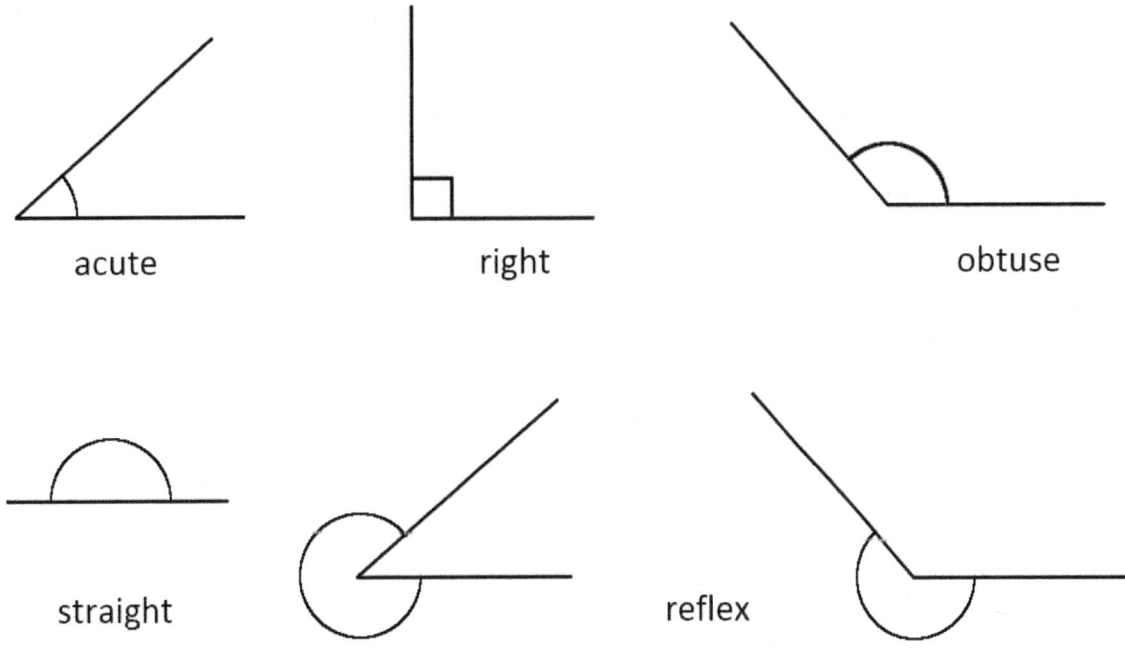

Exercise 20.1

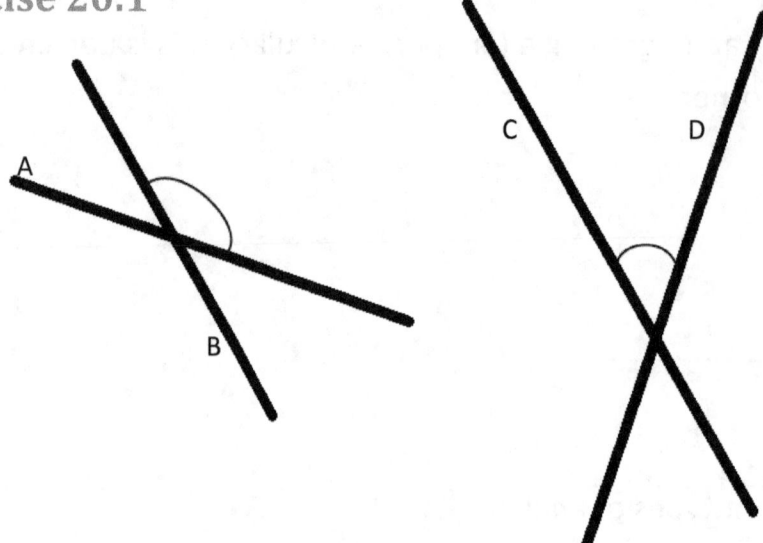

1. Which two lines are parallel? _____ and _____
2. Which two lines are perpendicular? _____ and _____
3. How many degrees in a straight line? _____
4. How many degrees in a right angle? _____
5. How many degrees in a full rotation (same as four right angles)? _____
6. Which angle depicted above is an acute angle (give the letters of the two lines)? _____
7. Which angle depicted above is an obtuse angle? _____

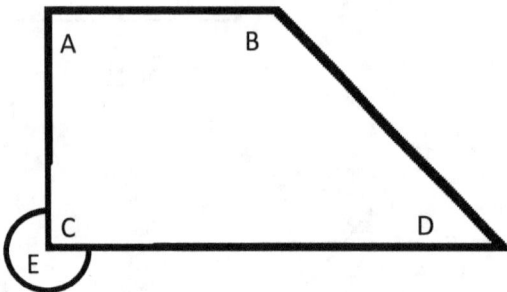

In the shape above, which angle is:

8. Acute? _____
9. Obtuse? _____
10. Reflex? _____

20.2 Lines and Angles

You need to know these angles:

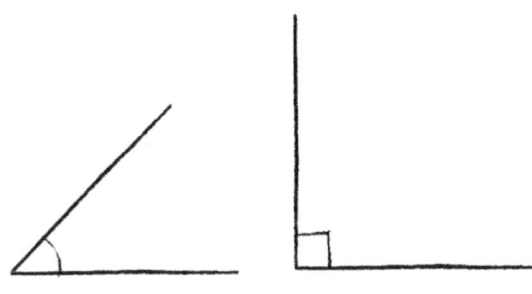
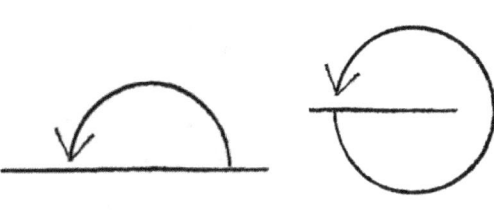

Half a right angle is 45°   There are 90° in a right angle   There are 180° in a straight line   There are 360° in a full rotation.

Complementary angles add up to 90°.

Supplementary angles add up to 180°.

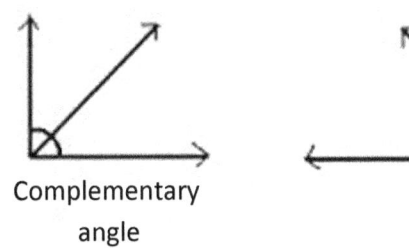

Complementary angle   Supplementary angle

Example:

Find angle **a**

As it is a straight line 150° + **a** must be 180°.

180°-150° = 30°.

So **a** is 30°.

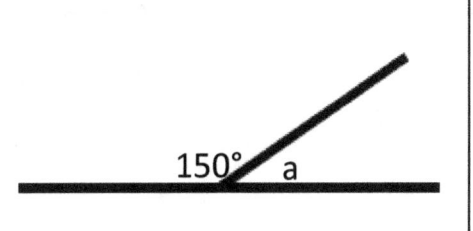

If two letters are the same then the angles are the same size.

Exercise 20.2

Calculate the missing angles below.

1.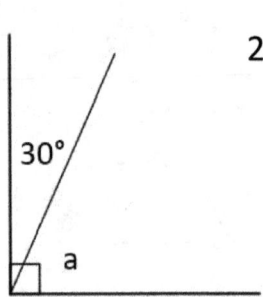
2.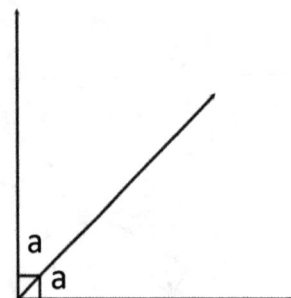
3.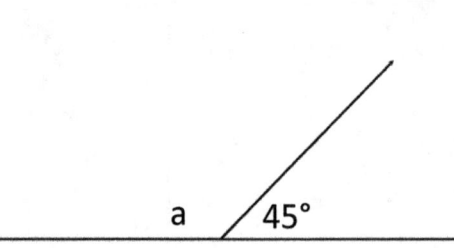

4.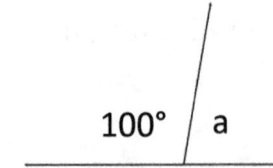
5.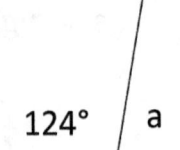
6.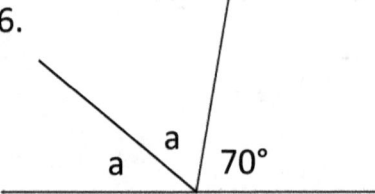

7.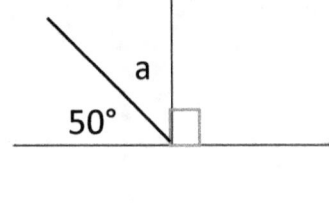
8.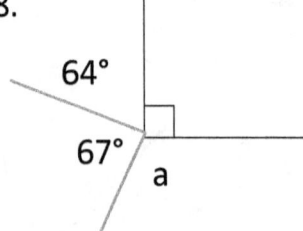
9.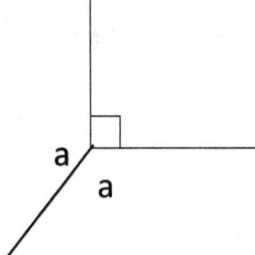

10.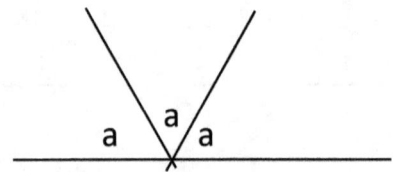

20.3 Alternate, Corresponding and Opposite Angles

When two lines intersect four angles are formed.

The opposite angles are equal.

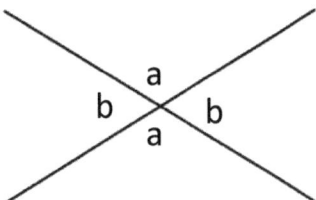

When a line intersects two parallel lines. Eight angles are formed.

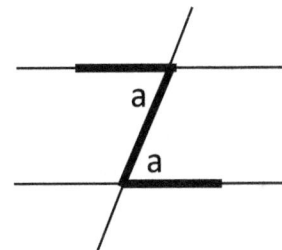

Alternate angles form a "Z" shape (in any direction).
Alternate angles are equal.

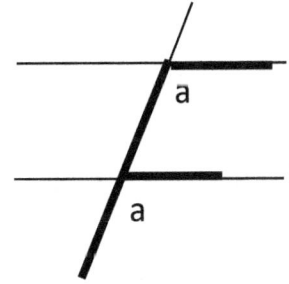

Corresponding angles form an "F" shape (in any direction).
Corresponding angles are equal.

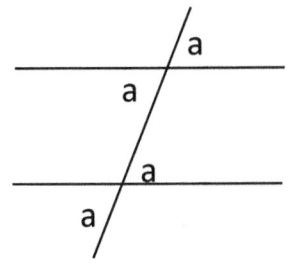

Vertically opposite angles are equal.

Exercise 20.3

Calculate the angles marked.

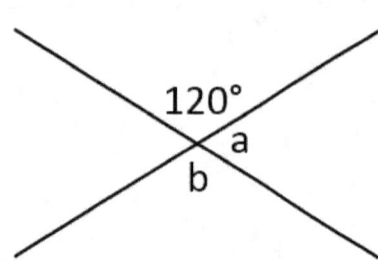

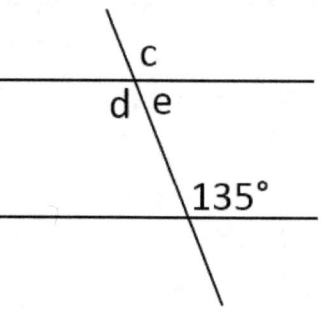

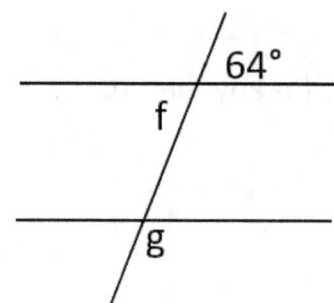

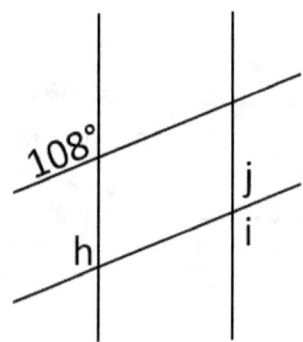

1. a° = _____

2. b° = _____

3. c° = _____

4. d° = _____

5. e° = _____

6. f° = _____

7. g° = _____

8. h° = _____

9. i° = _____

10. j° = _____

20.4 Triangles and Quadrilaterals

The interior angles of a triangle add up to 180°.

The interior angles of a quadrilaterals add up to 360°.

Exercise 20.4

1.

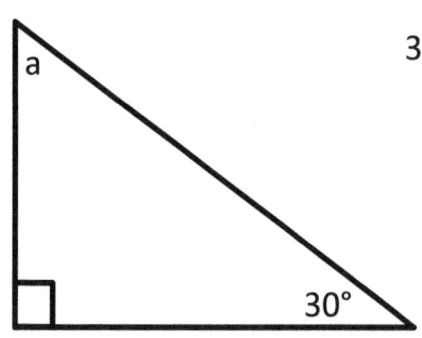

a° = _____

2.

a° = _____

3.

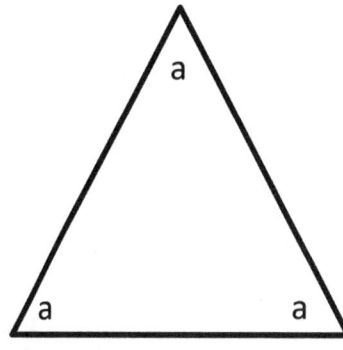

a° = _____

4.

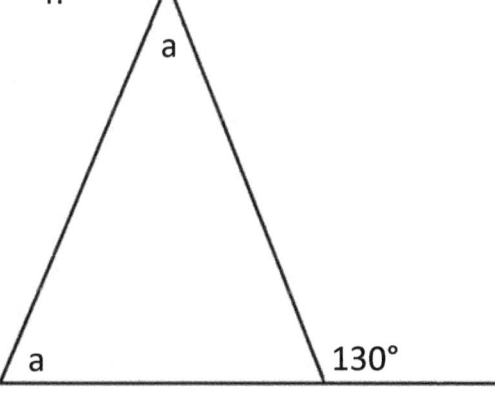

a° = _____

5.

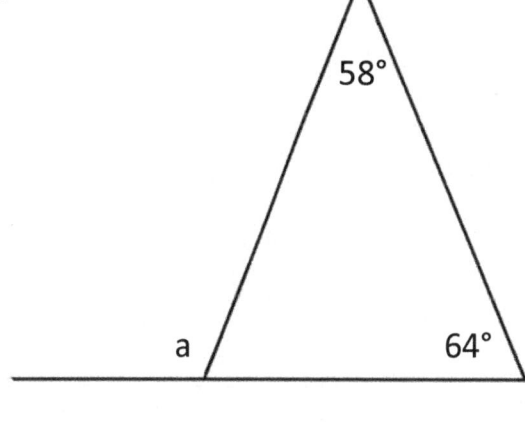

a° = _____

6.

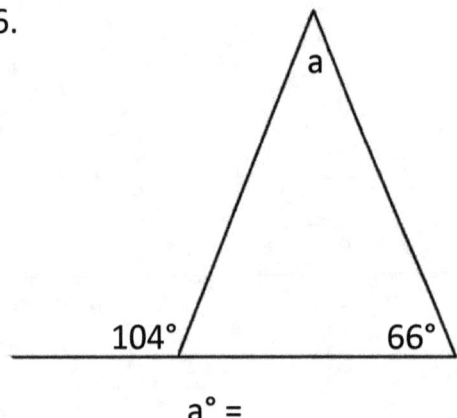

a° = _____

7.

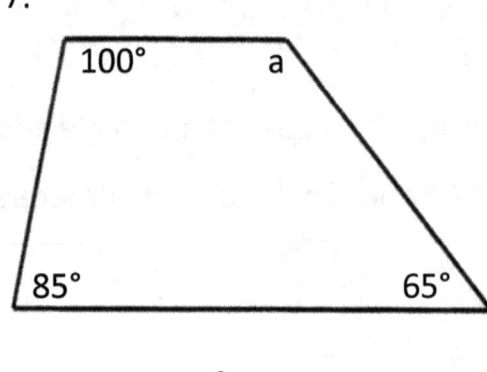

a° = _____

8.

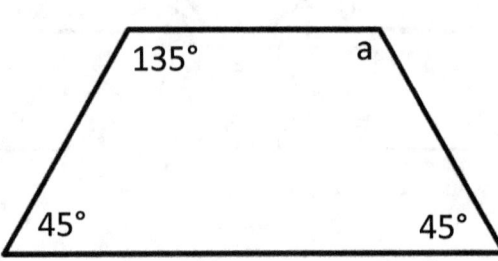

a° = _____

9.

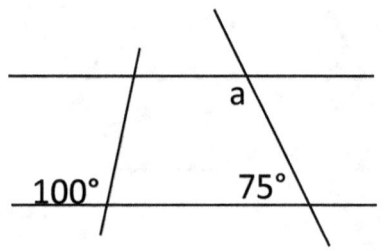

a° = _____

10.

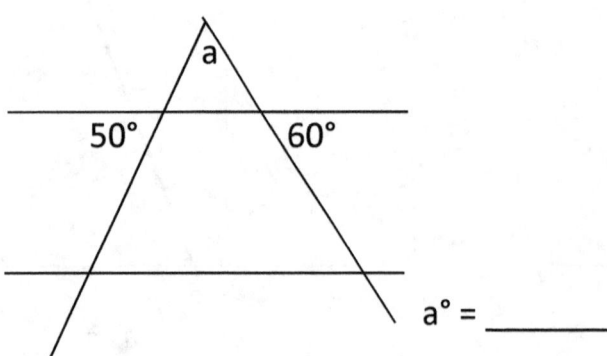

a° = _____

Chapter 21: 2D Shapes

21.1 Polygons

A polygon is a closed 2D shape with three or more straight sides.

The name of the polygon depends on the number of sides.

N° of sides	Name
3	Triangle
4	Quadrilateral
5	Pentagon
6	Hexagon
7	Heptagon
8	Octagon
9	Nonagon
10	Decagon

A regular polygon has equal sides *and* equal angles.

When drawing shapes, the same number of little lines across mean that the lines are the same length. The same number of arrows on the line represent that the lines are parallel.

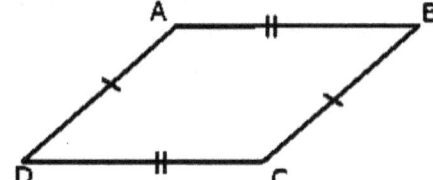 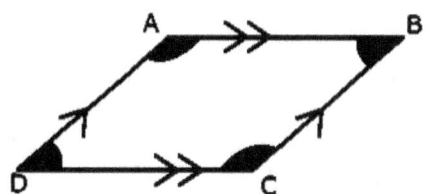

Showing which lines   Showing which lines
are the same length   are parallel.

Exercise 21.1

State whether the following shapes are regular or irregular and the name of the polygon.

1.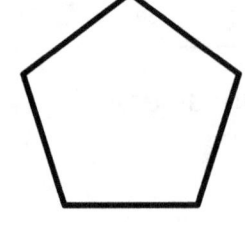
_____
_____

2.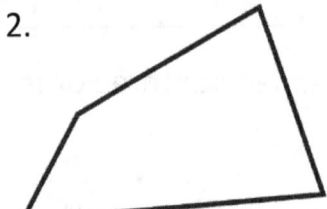
_____
_____

3.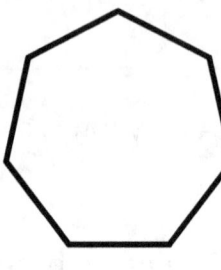
_____
_____

4.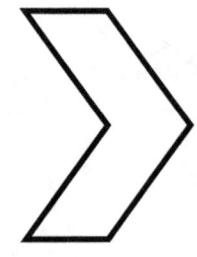
_____
_____

5.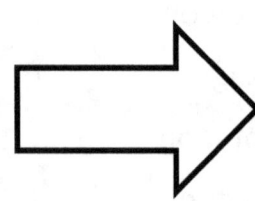
_____
_____

6.
_____
_____

7.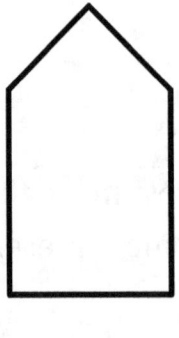
_____
_____

8.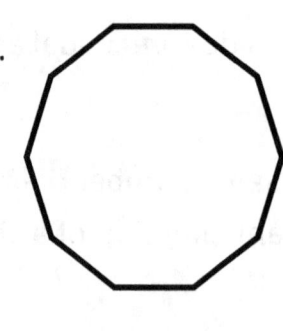
_____
_____

9.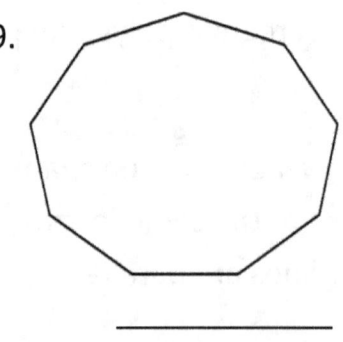
_____
_____

10.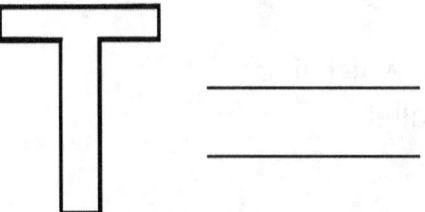
_____
_____

21.2 Triangles and Quadrilaterals

There are four types of triangles to know:

  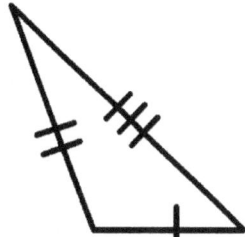 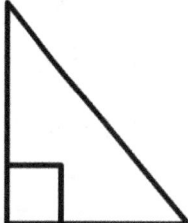

Equilateral

3 equal sides

3 equal angles

Isosceles

2 equal sides

2 equal angles

Scalene

No equal sides

No equal angles

Right-angle

Isosceles or scalene

One 90° angle

Quadrilaterals are polygons with four sides.

There are many types of quadrilaterals, including:

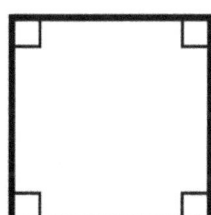 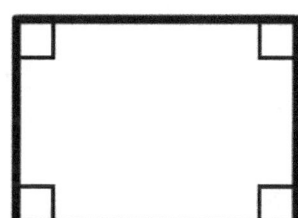

Square

4 equal sides

4 right angles

2 sets of parallel sides

Rectangle

2 pairs of equal sides

4 right angles

2 sets of parallel sides

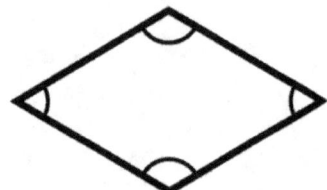

Rhombus

4 equal sides

2 pairs of equal angles

2 sets of parallel sides

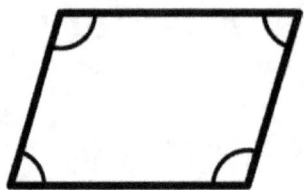

Parallelogram

2 pairs of equal sides

2 pairs of equal angles

2 sets of parallel sides

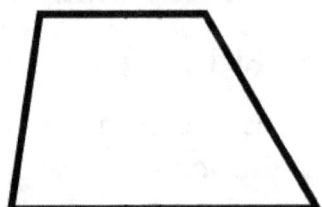

Trapezium

1 set of parallel sides (not equal)

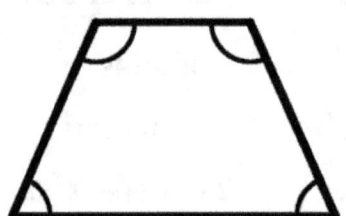

Isosceles trapezium

1 pair of equal sides

2 pairs of equal angles

1 set of parallel sides (not equal)

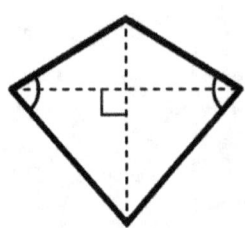

Kite

2 pairs of equal sides

1 pair of equal angles (obtuse)

No parallel sides

No internal reflex angles

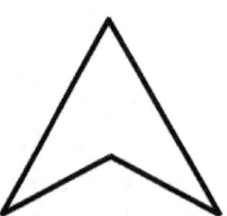

Arrowhead

2 pairs of equal sides

1 pair of equal angles (acute)

No parallel sides

1 internal reflex angle

Exercise 21.2

1. What is a regular quadrilateral called? _____

2. What size is an angle in an equilateral tringle? _____

3. Two identical equilateral triangles are placed together with one side touching. What is the name of the shape that is formed? _____

4. In an isosceles trapezium, if one angle is 120°, what are the sizes of the other three angles? _____, _____, _____

5. Both kites and arrowheads have one pair of equal angles. How are the angles different in the two shapes? _____
   _____

6. Complete the analogy:
   as square is to rectangle so rhombus is to _____

7. Parallelograms and trapeziums each have two pairs of equal angles. In both shapes, one set is always obtuse and one set acute. How are the angles different in the two shapes? _____
   _____

Name the shapes below:

8.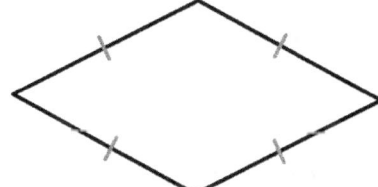

9.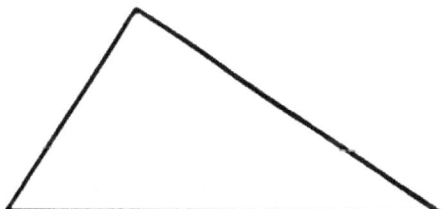

   _____   _____

10. Name all the different types of quadrilaterals you can find in this regular hexagon. _____
    _____
    _____

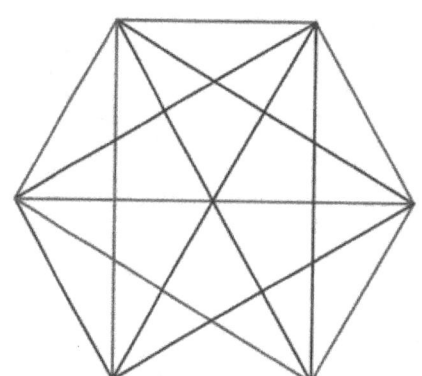

21.3 Circles and Ellipses

A circle is a set of points all an equal distance from the centre.

Some circle words are:

Circumference – the distance around the circle. (The perimeter of the circle).

Arc – a small part of the circle.

Radius (plural: radii) – the distance from the centre of the circle to the outside.

Diameter – the distance from one side of the circle to the other, going through the centre.

The diameter = 2 x radius

Quadrant = a quarter of a circle.

Semi-circle – half a circle.

Sector – an area of the circle between two radii, that is not a semi-circle or quadrant.

Chord – line going from one part of the circle to another, without going through the centre.

Segment – part of a circle between a chord and the outside of the circle.

An ellipse or oval is another curved shape and is *not* a circle.

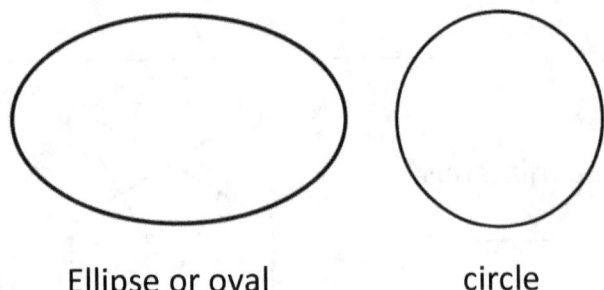

Ellipse or oval   circle

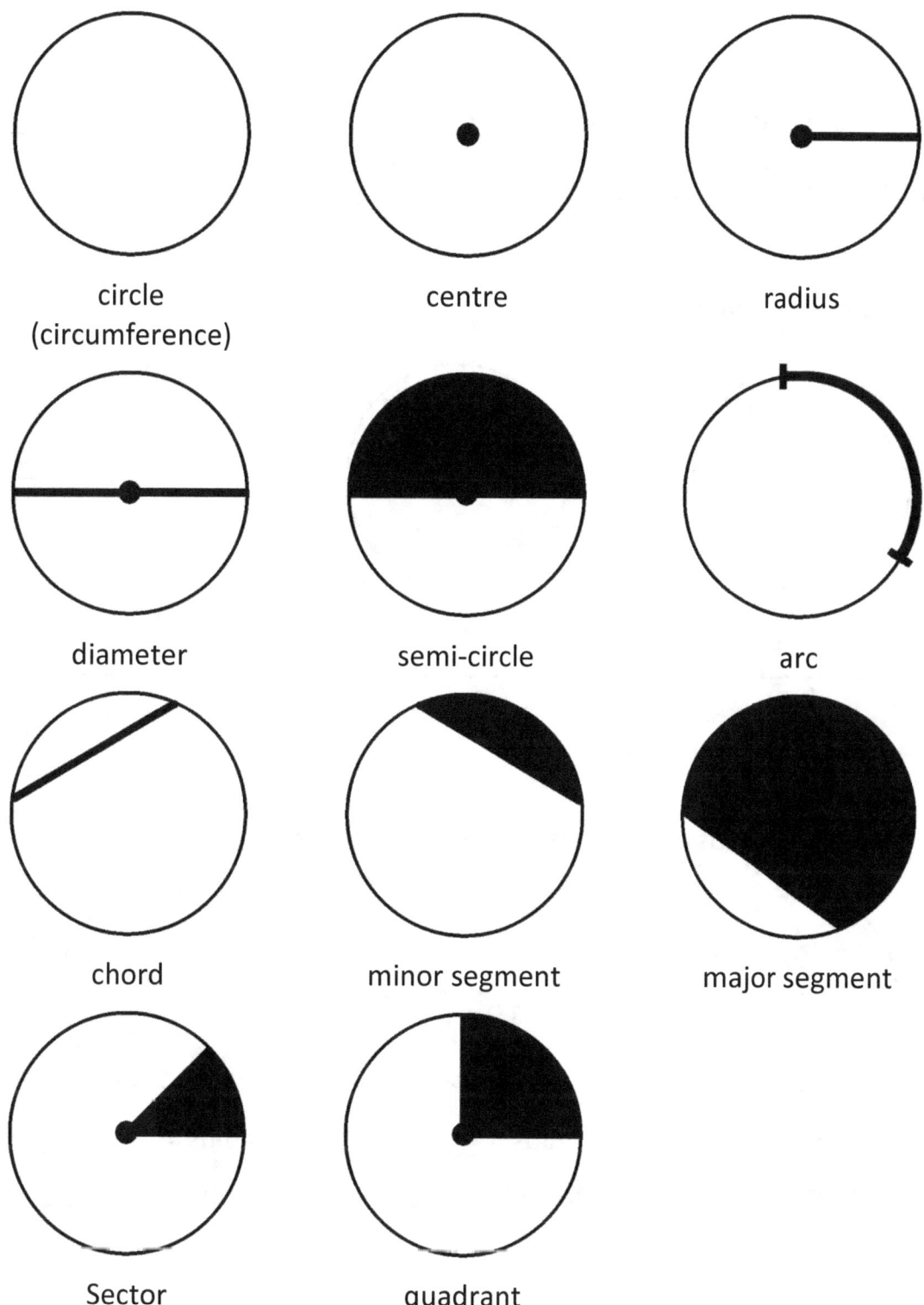

Exercise 21.3

1. What is the perimeter of a circle called? _____

2. What is half a circle called? _____

Label the following:

3. 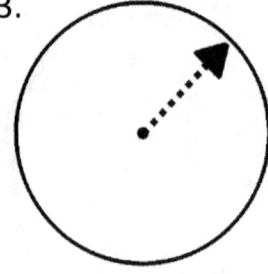   4. 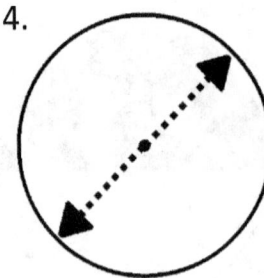   5.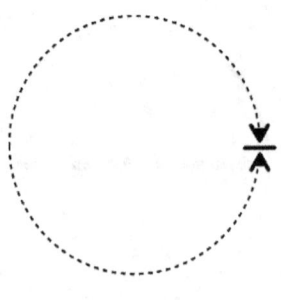

_____   _____   _____

What is the radius and diameter of the circles below?

   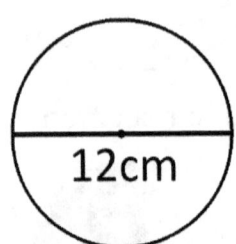   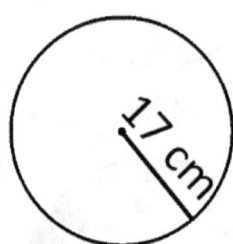

radius = _____   radius = _____   radius = _____

diameter = _____   diameter = _____   diameter = _____

9. Why is 'a' not the radius? 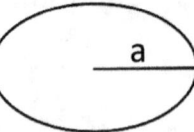 _____

10. What is the difference between a diameter and a chord? _____
_____

Chapter 22: Area and Perimeter

22.1 Perimeter

The perimeter is the distance around the outside of a 2D shape.

To work out the perimeter add up the lengths of all the sides.

Example 1: What is the perimeter of the shape below?

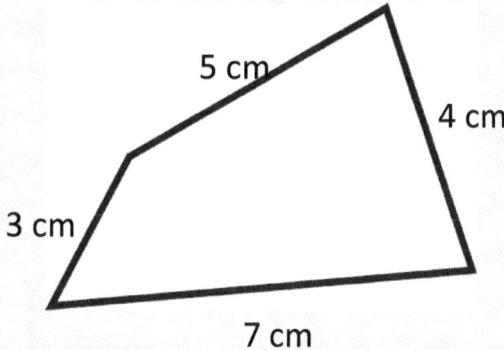

Perimeter = 3+4+5+7 = 19cm

Example 2: What is the perimeter of a regular pentagon with a side of 4cm?

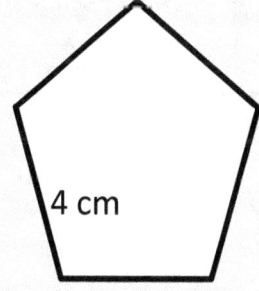

Perimeter = 4 + 4 + 4 + 4 + 4
= 5 x 4
= 20 cm.

Diagrams are generally *not* drawn to scale.

Exercise 22.1

Calculate the perimeter of the shapes below.

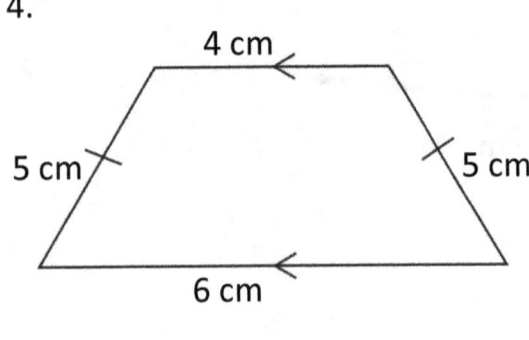

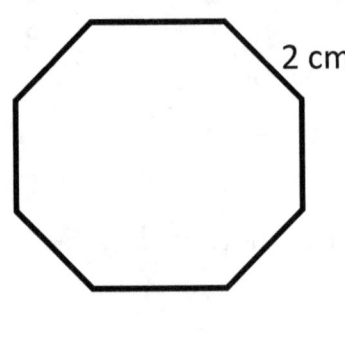

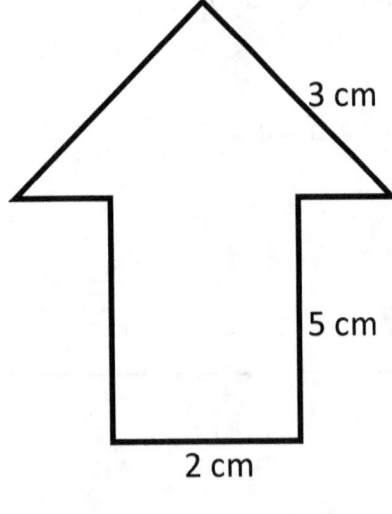

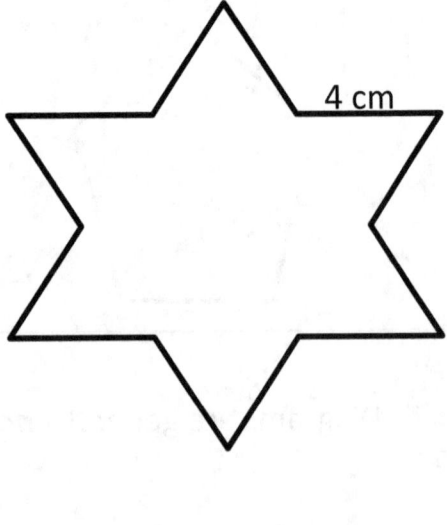

8.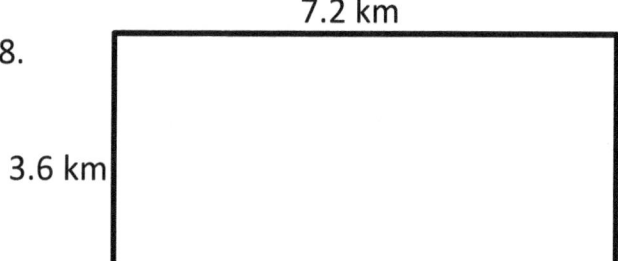

9.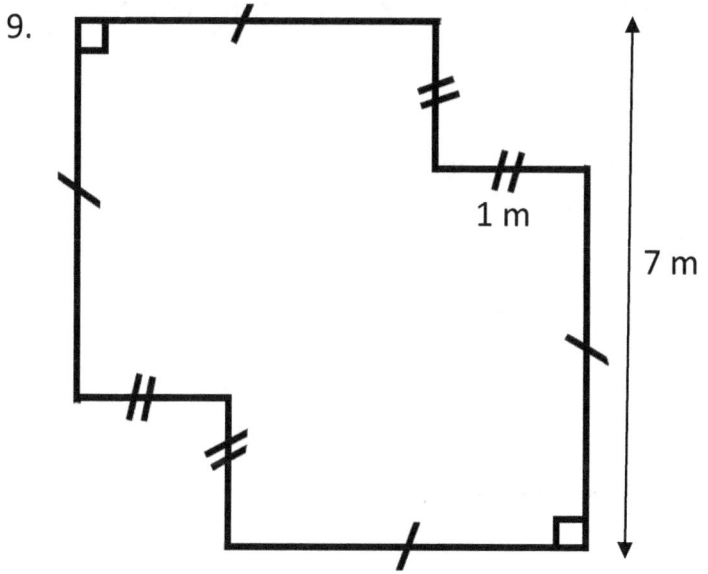

10.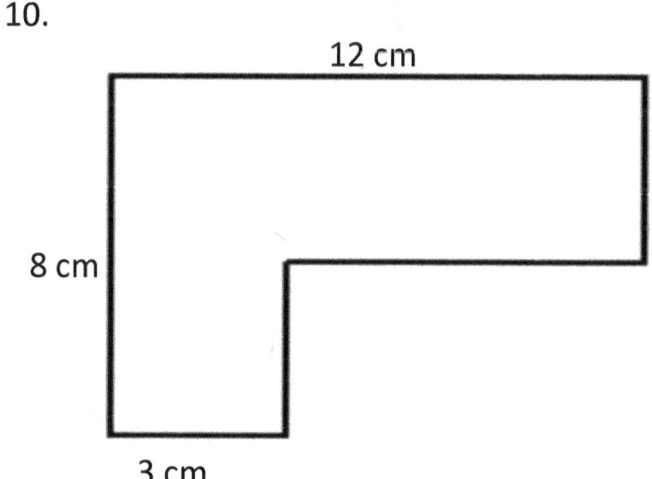

22.2 Area of a rectangle

The area of a rectangle can be calculated by the formula:

Area = length x width

This can be written as:

A = l x w

The units for area are the units used for length squared, eg. mm^2, cm^2 or m^2.

For a square, since length and width are the same:

A (square) = l^2

Example:

[rectangle: 4 cm by 2 cm]

Area = l x w
= 4 x 2
= 8 cm^2

Exercise 22.2

Find the area of the following rectangles. Remember the units.

1.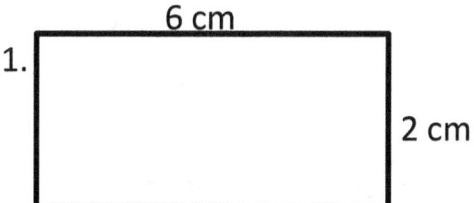

2.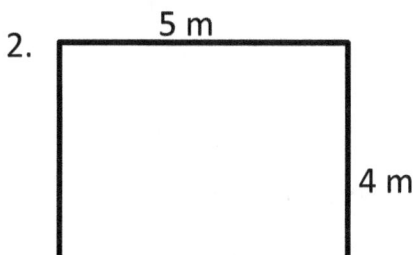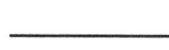

3.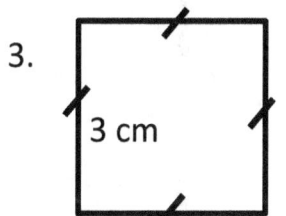

4.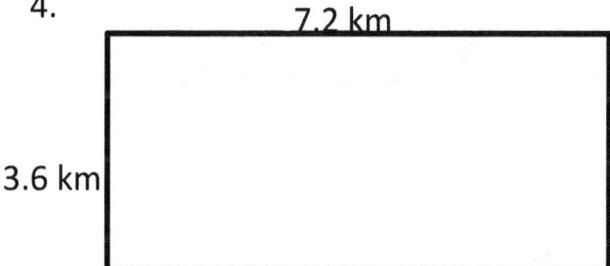

5.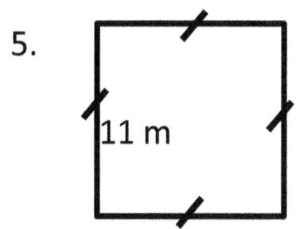

6. A house is surrounded by a fence, with a length of 60m and a width of 25m. What is the area of the property? _____

7. A swimming pool has a length of 25m and had eight lanes each with a width of 1.25m. What is the area that the pool takes up? _____

8. A volleyball court has the dimensions 9m by 18m. What area does a volleyball court take up? _____

Questions 9 and 10 refer to the diagram of the netball court below.

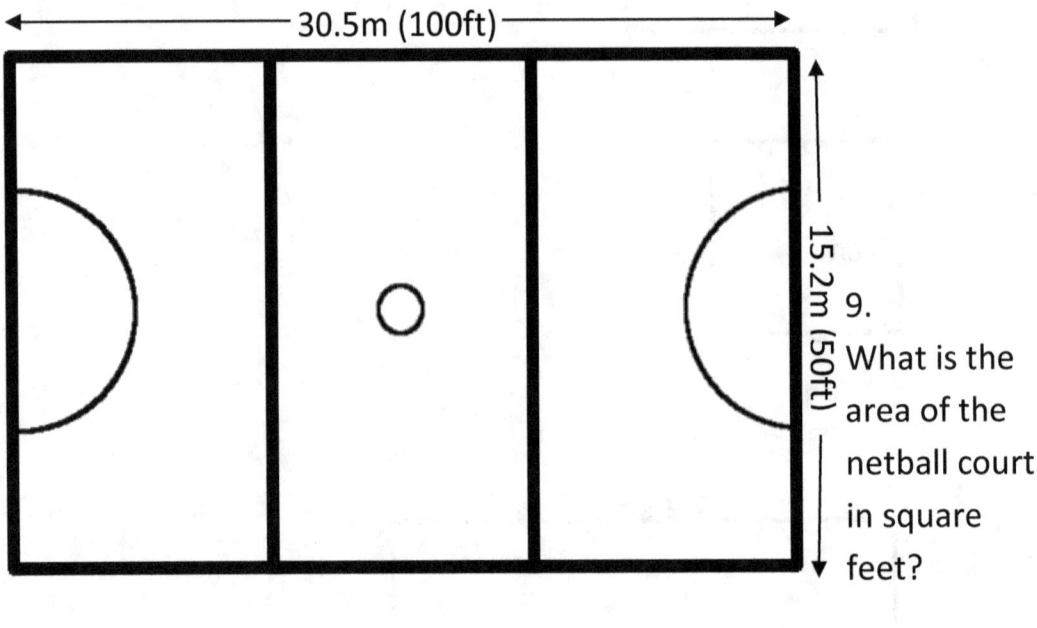

9. What is the area of the netball court in square feet? _____

10. What is the area of the netball court in metres squared? _____

22.3 Area of a triangle

The area of a triangle can be calculated by the formula:

$$\text{Area} = \frac{1}{2} \text{ base} \times \text{height}$$

This can also be written as:

$$A = \frac{1}{2} b \times h \qquad \text{or} \qquad A = \frac{1}{2} bh$$

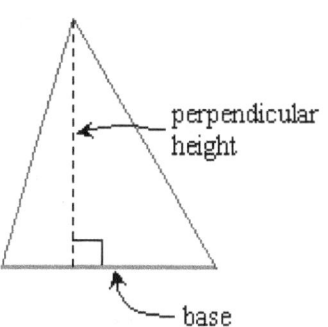

The height is the perpendicular height, only in a right-angled triangle will that be the same as a side. In a triangle with an obtuse angle, this will be outside the triangle.

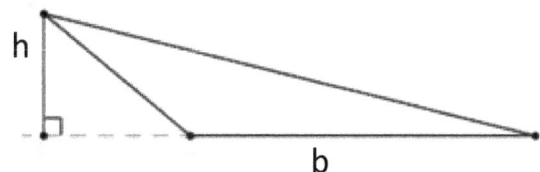

Example: Find the area of the triangle below.

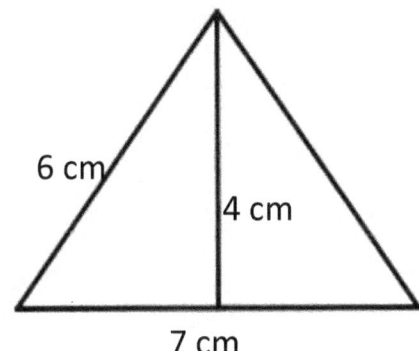

$$A \text{ (triangle)} = \frac{1}{2} b \times h$$

$$= \frac{1}{2} \times 7 \times 4$$

$$= 14 \text{ cm}^2$$

Remember: as it makes no difference what order numbers are multiplied in either number may be halved.

Exercise 22.3

Calculate the area of the triangles below. Remember the units.

1.

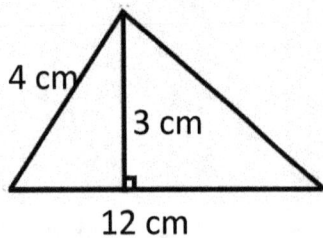

2.

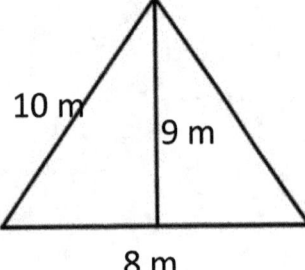

3.

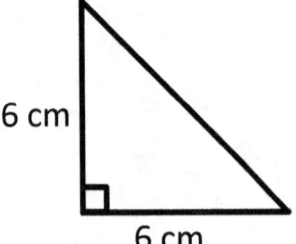

4.

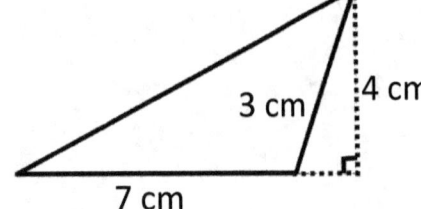

5.

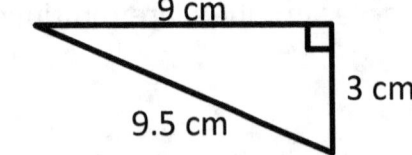

6.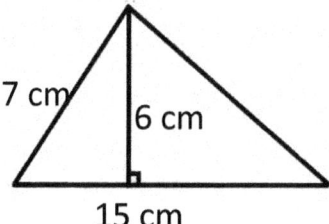

7.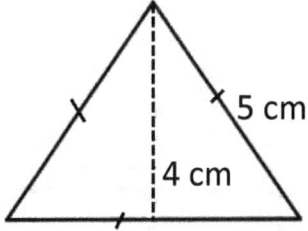

8.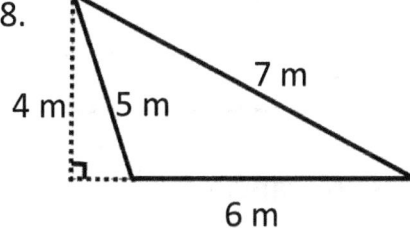

9.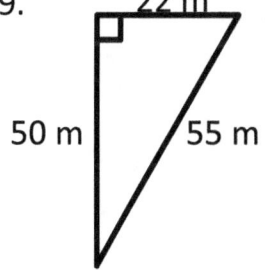

10.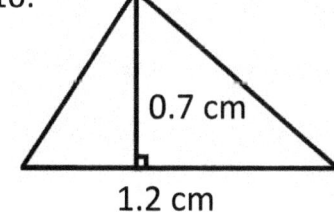

22.4 Complex Area

To find the area of a more complex shape:

- Divide the shape into rectangles and/or triangles.
- Work out any missing dimensions.
- Find the area of each section individually.
- Add or subtract the areas as necessary.

Example: A rectangular park (30m x 24m) is surrounded by a path of width 2m. What is the area of the path?

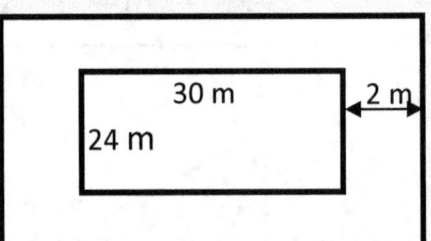

Answer: First work out the dimensions of the larger rectangle.

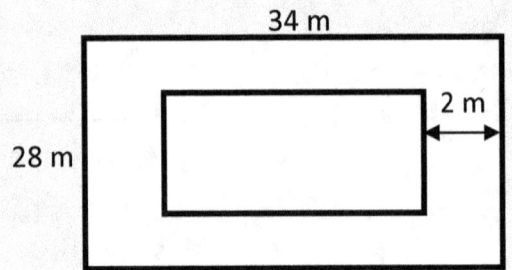

A (large) = l x w

= 34 x 28

= 952 m²

A (small) = l x w

= 30 x 24

= 720 m²

A (path) = A (large) – A (small)

= 952 – 720

= 232 m²

Example: Find the area of the shape below.

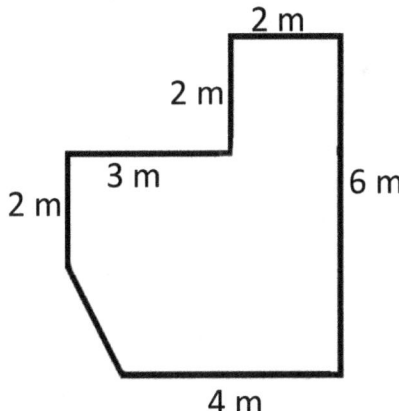

First divide into rectangles and triangles.

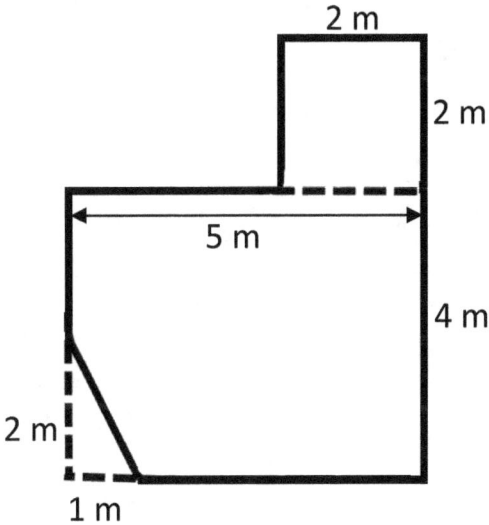

Work out individual areas:

Area (square) = $l^2 = 2^2$
            = 4 m^2

Area (rectangle) = l x w

   = 4 x 5

   = 20m^2

Area (triangle) = $\frac{1}{2}$ b x h

   = $\frac{1}{2}$ 1 x 2

   = 1m^2

Area (total) = A (square) + A (rectangle) − A(triangle)

   = 4 + 20 − 1

   = 23 m^2

Exercise 22.4

1. What is the area of this shape?

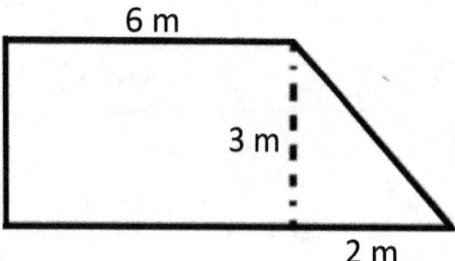

2. What is the area of this shape?

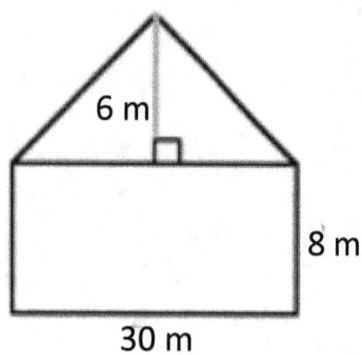

3. What is the area of this shape?

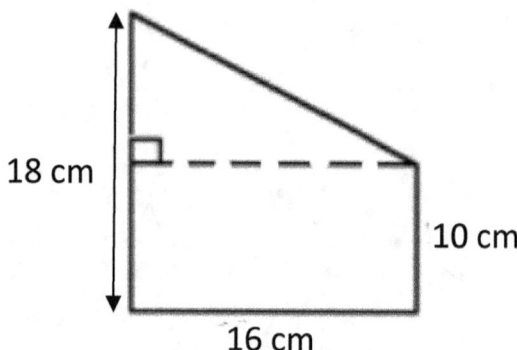

4. How much carpet is needed to carpet this "L" shaped room?

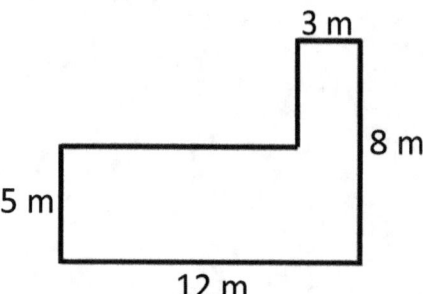

5. A path was laid around a garden. What is the area of the path? _____

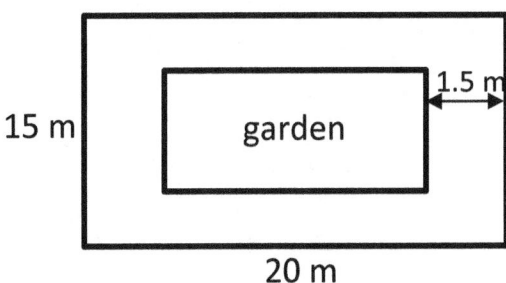

6. Annie wants to paint the front of an arch way, but needs to know the area, to work out how much paint to buy. What is the area of the front of the arch? _____

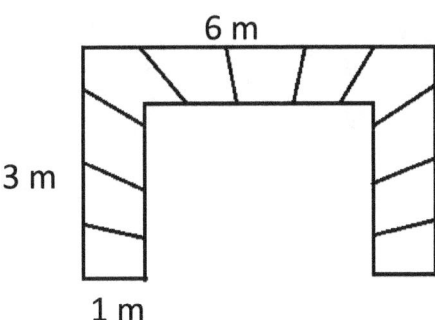

7. A decorator decides to put a layer of square tiles around the outside of this mirror (so the tiles make a frame which the mirror fits into). The tiles are 25cm^2. How many tiles will he need? _____

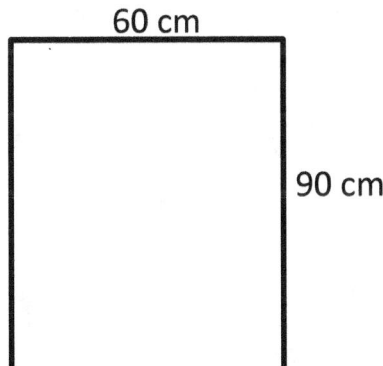

8. What is the total area needed to hang the mirror (question 7) with the tile border? _____ cm^2

9. What is the area of the trapezium below? _____

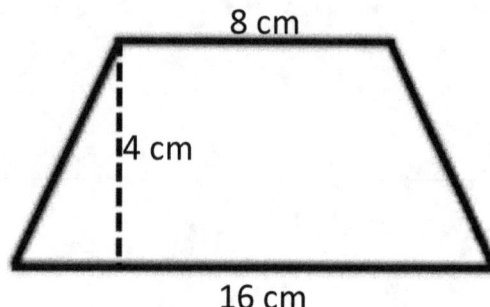

The standard method of calculating the area of a trapezium is to average the two parallel sides (base and top) and multiply by the perpendicular height. Have a go at using that formula below. Did you get the same answer? _____

10. What is the area of the regular hexagon below? _____

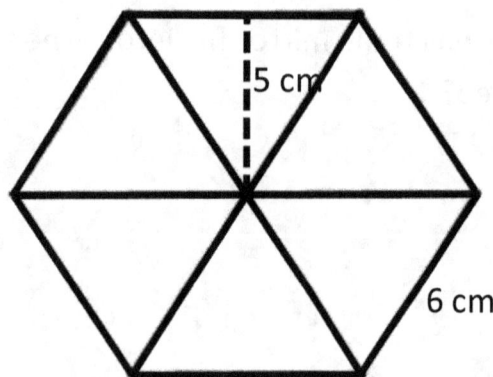

Chapter 23: 2D Shape Problems

23.1 Reverse Area and Perimeter Problems

If you are given all the measurements except one and either the area or the perimeter, you can work out the other measurement.

To do this just remember that whatever you do to one side of an equation, to the other side you must do **exactly the same**.

Example: What is the length of the rectangle below.

Area = 54 cm², 6 cm

$$\text{Area} = l \times w \quad \text{write the equation}$$
$$54 = l \times 6 \quad \text{put in what you know}$$
$$\frac{54}{6} = \frac{l \times 6}{6} \quad \text{To get just the letter (l)}$$
$$9 = l \quad \text{on its own, divide each side by 6.}$$

So, the missing length is 9cm.

Example: What is the missing length in the shape below?

7 cm, 3 cm, 5 cm, Perimeter = 24cm

$$\text{Perimeter} = s1 + s2 + s3 + s4$$
$$24 = 5 + 7 + 3 + l$$
$$24 = 15 + l$$
$$24 - 15 = l$$
$$9 = l$$

So, the missing length is 9cm.

Exercise 23.1

Calculate the missing lengths of the shapes below.

1. Area = 49cm²   _____

2. Area = 180cm², 12 cm   _____

3. Perimeter = 70cm, 15cm   _____

4. Perimeter = 50cm, 3.5cm, ?, 14cm   _____

5. The perimeter of this regular octagon is 96cm.   _____

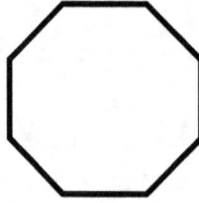

6. 8cm, Area = 56cm²   _____

7.

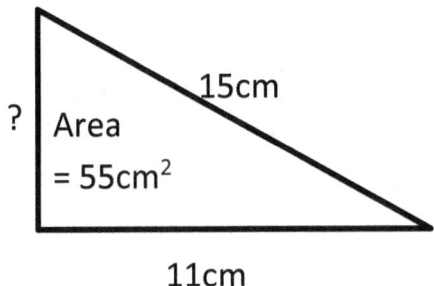

8. The area of the triangle below is 48cm². What is the length of the base?

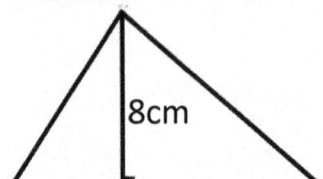

9. The area of a triangle is 42cm². If the base is 7cm, what is the perpendicular height of the triangle?

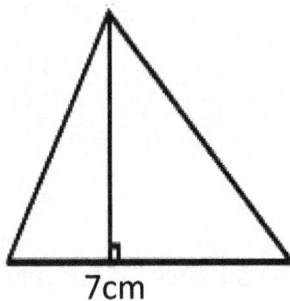

10. What is the length of the side marked "?" The total area of the shape is 47.5cm².

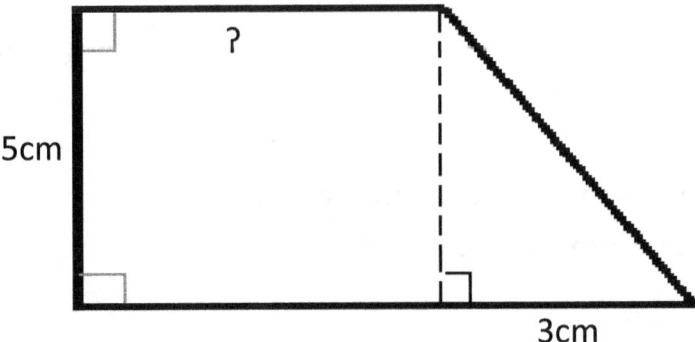

23.2 Area and perimeter

The area of a shape can be used to find the perimeter and vice versa. Simply, look at the question and work out what you need to answer it. You can then find any missing information by using the area or perimeter given.

Example:

What is the perimeter of the rectangle?

12cm, Area = 60cm²

For perimeter I need the width of the rectangle.

Area = l x w
60 = 12 x w
$\frac{60}{12} = \frac{12 \times w}{12}$
w = 5 cm

Perimeter = (2 x l) + (2 x w)
= (2 x 12) + (2 x 5)
= 24 + 10
= 34cm

Exercise 23.2

1. Area = 144cm² What is the perimeter of a square that has an area of 144cm²? _____

2. Area = 9cm² What is the perimeter of a square that has an area of 9cm²? _____

3. Perimeter = 20cm What is the area of a square that has a perimeter of 20cm? _____

4. 15cm — Perimeter = 40cm

What is the area of this rectangle, which has a perimeter of 40cm? _____

5. 2cm | Area = 22cm²

What is the perimeter of this rectangle which has an area of 22cm²? _____

6. 5cm, 3cm

What is the area of this triangle if it has a perimeter of 12cm? _____

7. 5cm, 7cm

What is the perimeter of this triangle if it has an area of 12.5cm²? _____

8. 3.1cm

What is the area of this rectangle if it has a perimeter of 8.8cm? _____

9.

What is the area of a square if the perimeter is 8.4cm? _____

10. 3cm

What is the area of this triangle if it has a perimeter of 15cm? _____

23.3 Reflective Symmetry

A line of reflection can be thought of in three ways

- A mirror line.

- A fold line - when folded along the line the two halves are on top of each other.

- A line where every point one side of the line has a corresponding point the same distance from the line on the other side.

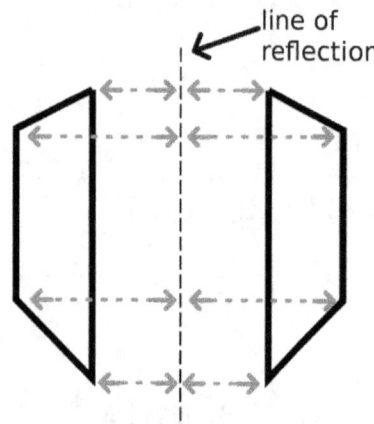

All of these points are the same distance from the line of reflection.

For a regular polygon, the number of sides equals the number of lines of reflection.

Exercise 23.3

How many lines of symmetry in the shapes below?

1.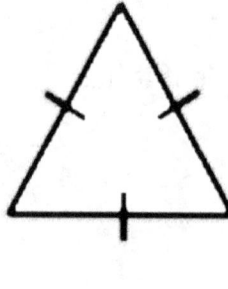

2.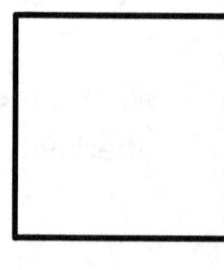

_____   _____

3.

4.

5.

6.

7.

8.

9.

10.

23.4 Rotational Symmetry

A shape has Rotational Symmetry when it looks the same after a rotation (of less than one full turn).

Looking at a six pointed star. One point has been shaded to show the starting point.

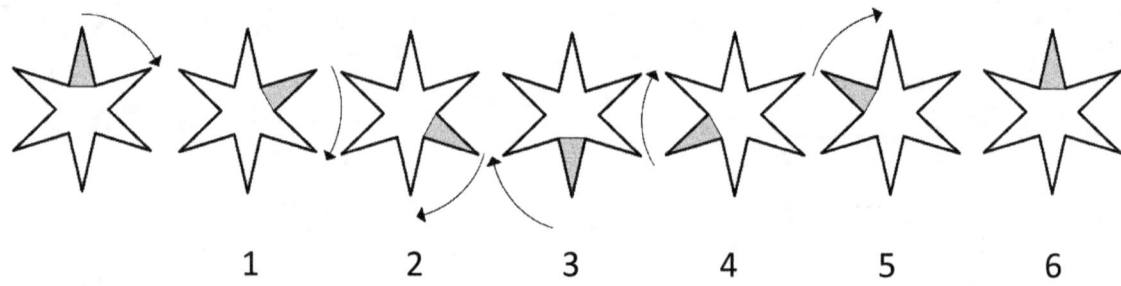

1 2 3 4 5 6

So, the star has a rotational order of 6, because it looks exactly the same as the starting point, six times in a full rotation.

For a regular polygon, the number of sides is the same as the order of rotational symmetry.

A shape with an order of symmetry of one is said to have *no* rotational symmetry. This is because there is only one position that the shape looks a particular way.

Exercise 23.4

What order of rotational symmetry do the shapes have below?

1.

2.

_____ _____

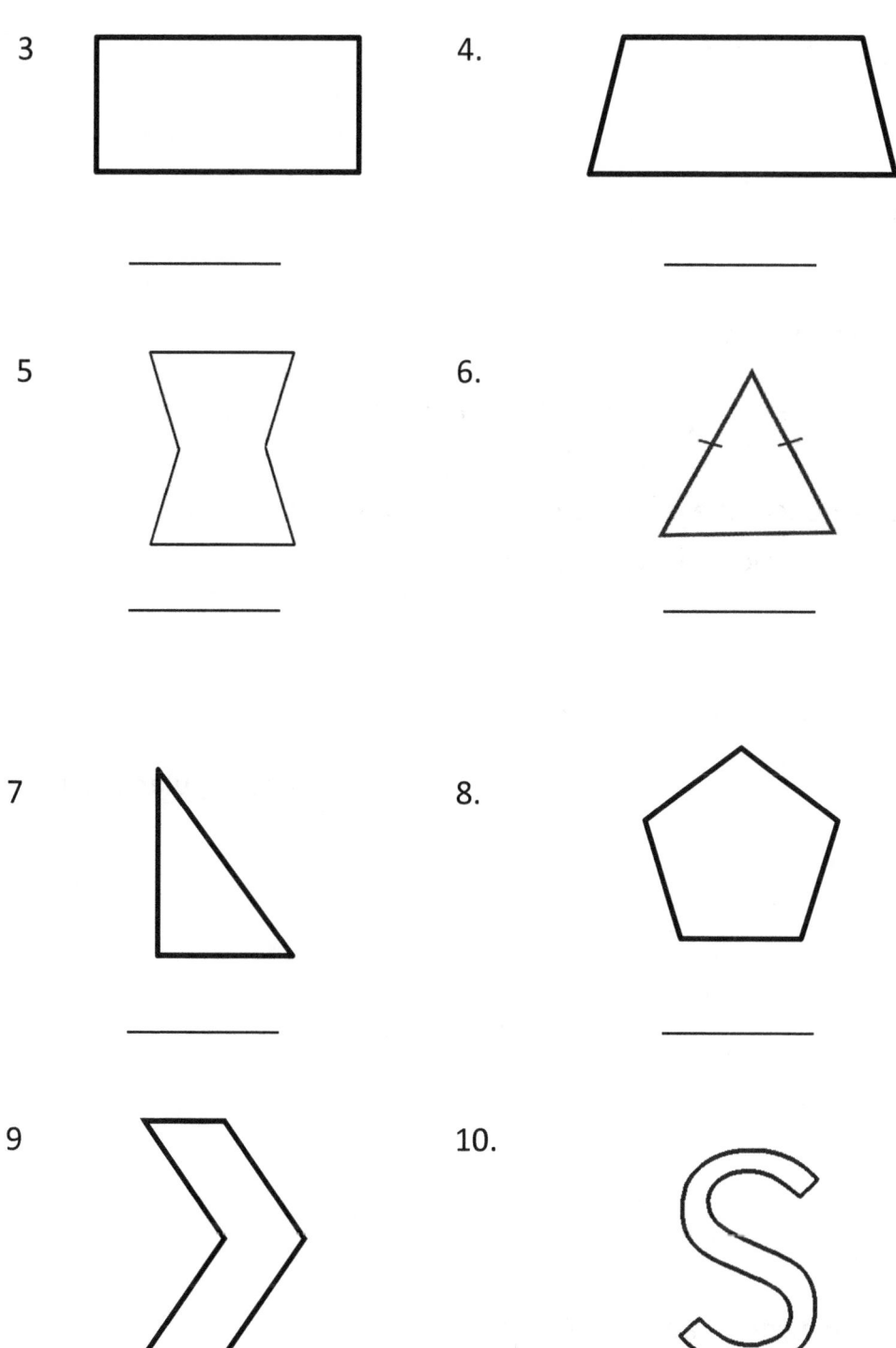

Chapter 24: 3D Shapes

24.1 Prisms and Pyramids

A prism is a 3D shape that:

- Has identical ends – they are identical to each other, each one is called a base.
- Flat sides – each side is a parallelogram (often rectangles).
- Same cross-section all along its length.

Prisms are named according to the shape of their ends.

For example: 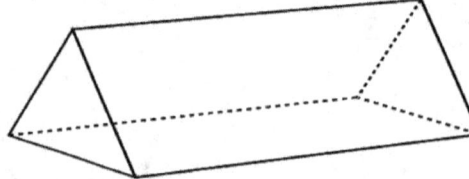 is a triangular prism.

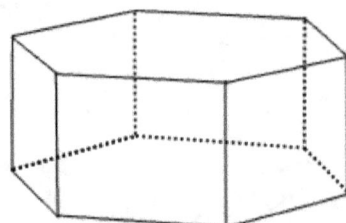 is a hexagonal prism.

Some special prims you need to know are:

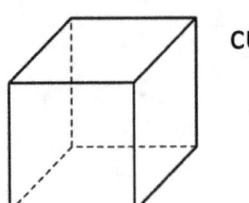 cube   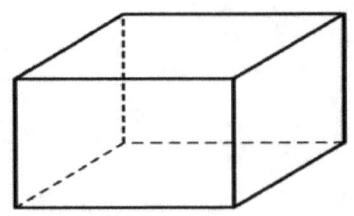 cuboid

A trapezium based prism can also be called a trapezoid.

A pyramid is a 3D shape that:

- The base is a polygon.
- The sides are triangles that meet at a point.

Pyramids are named according to the shape of their base.

For example:

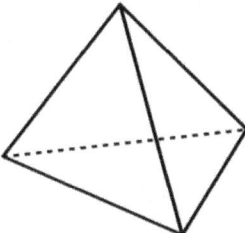 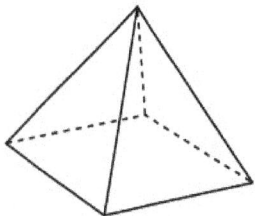

is a triangular pyramid. is a square-based pyramid.

A triangular pyramid is also called a tetrahedron.

An octahedron is a shape with eight triangles.

Curved 3D shapes

There are three curved 3D shapes that you need to know:

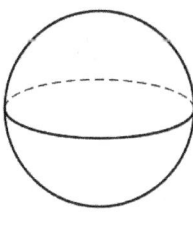  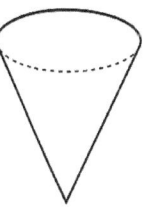

sphere cylinder cone

Technically a cylinder is not a prism and a cone is not a pyramid, because they have curved sides. However, they are very similar and often referred to as prism-like and pyramid-like respectively.

All 3D shapes have length, width and height.

They have a number of:

- Faces – a flat side.
- Edges – a line which two faces touch.
- Vertices (singular vertex) – the corners, where three or more faces meet.

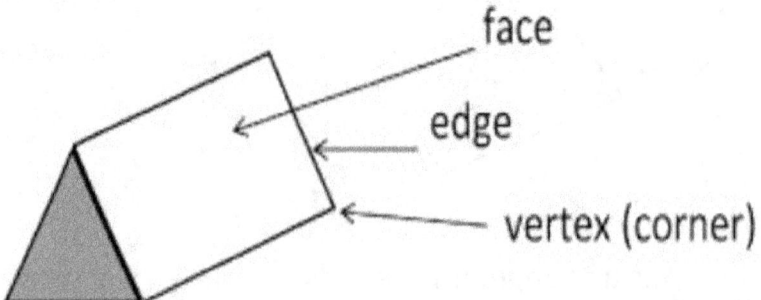

Exercise 24.1

For the following shapes give their name, the number of faces, edges and vertices.

1.
2.
3.

Name: _____ _____ _____

N° of faces: _____ _____ _____
N° of edges: _____ _____ _____
N° of vertices: _____ _____ _____

4.

Name: _____

N° of faces: _____
N° of edges: _____
N° of vertices: _____

5.

Name: _____

N° of faces: _____
N° of edges: _____
N° of vertices: _____

6.

Name: _____

N° of faces: _____
N° of edges: _____
N° of vertices: _____

7.

Name: _____

N° of faces: _____
N° of edges: _____
N° of vertices: _____

8.

Name: _____

N° of faces: _____
N° of edges: _____
N° of vertices: _____

9.

Name: _____

N° of faces: _____
N° of edges: _____
N° of vertices: _____

10.

Name: _____

N° of faces: _____
N° of edges: _____
N° of vertices: _____

24.2 Volume

The volume of a prism is the area of the base times the height.

$$\text{Volume (prism)} = \text{Area of base} \times h$$

This means that the volume for a cuboid is:

$$\text{Volume (cuboid)} = l \times w \times h$$

In a cube where length, width and height are the same number:

$$\text{Volume (cube)} = l^3$$

The units for volume is the units for distance cubed, eg. cm^3 and m^3.

Example: What is the volume of the cuboid below:

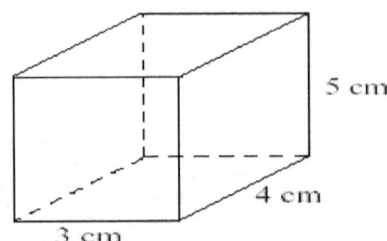

$$\text{Volume (cuboid)} = l \times w \times h$$
$$= 4 \times 5 \times 3$$
$$= 60 \text{ cm}^3$$

Volume of a pyramid is one third the volume of the prism:

$$\text{Volume of pyramid} = \frac{1}{3} \text{ area of base} \times h$$

Exercise 24.2

What is the volume of the shapes below?

1. 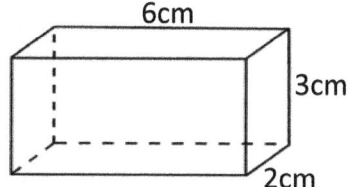   _____

2. 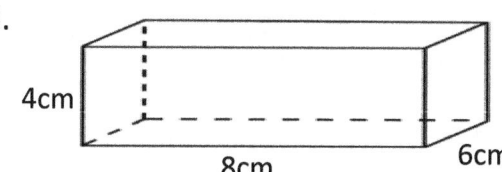   _____

3. 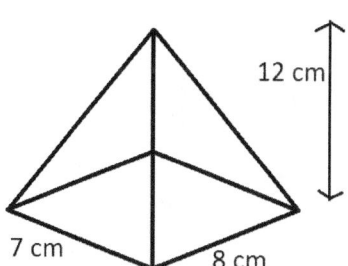   _____

4. 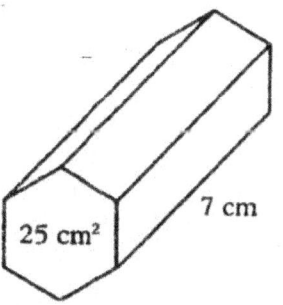   _____

5. 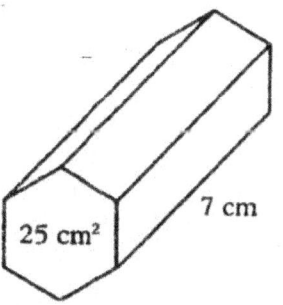   _____

6. 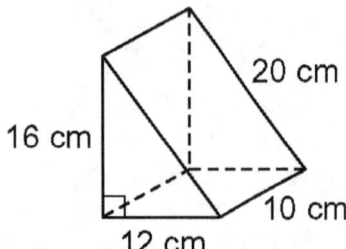 _____

7. 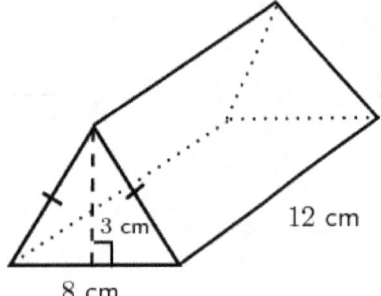 _____

8. 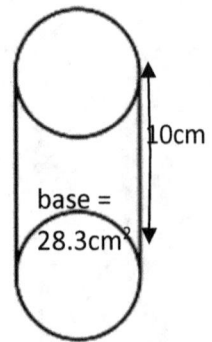 _____

9. 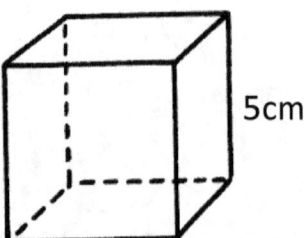 _____

10. 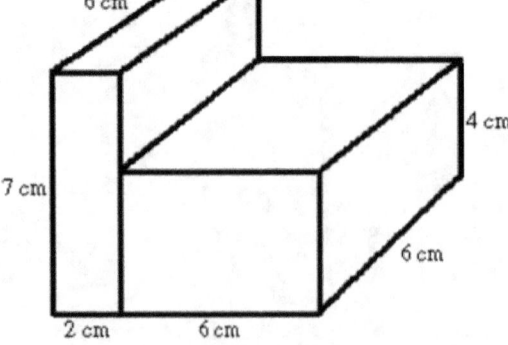 _____

24.3 Surface Area

To calculate the surface area of a 3D shape is the area of all the faces added together.

To calculate the surface area of a 3D shape with equal faces, calculate the area of one face and multiply by the number of faces.

Exercise 24.3

1. 5cm _____

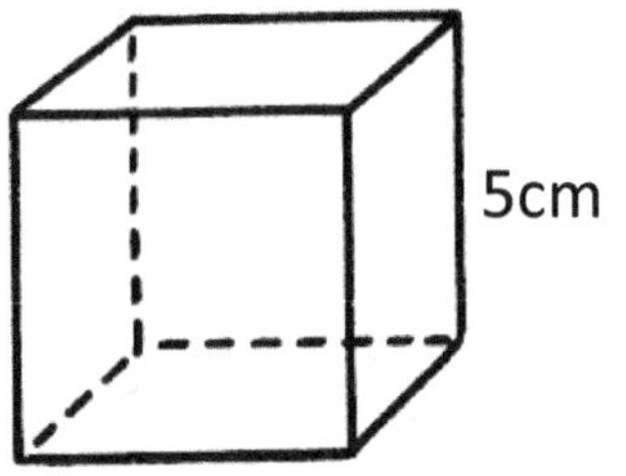

2. 2 cm _____

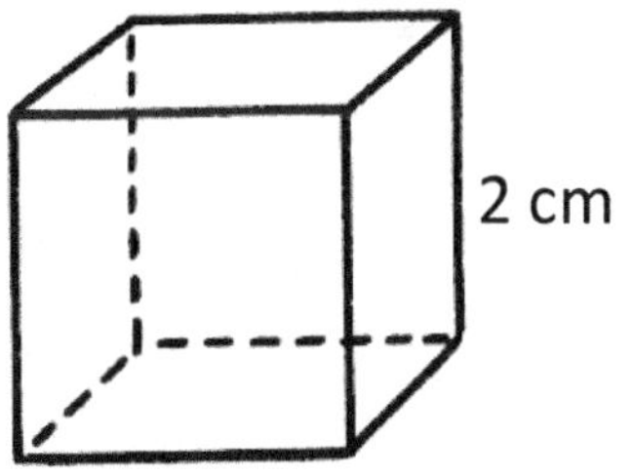

3. 1.5 cm _____

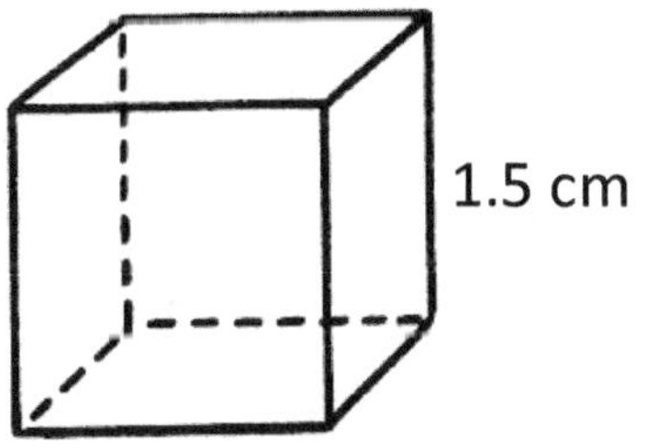

4. 8cm, 6cm, 6cm _____

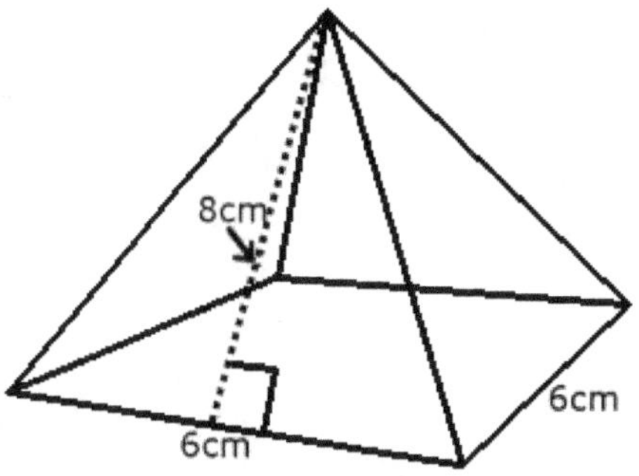

5. 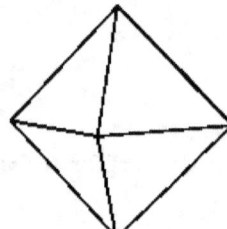 Octahedron, each side = 7cm² _____

6. 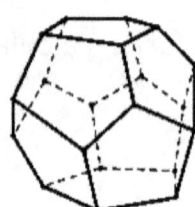 Dodecahedron (12 faces) _____

   Each side = 6cm²

7. 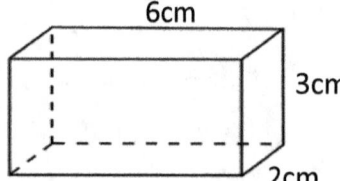 _____

8. 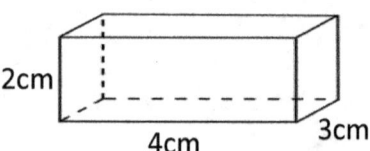 _____

9. 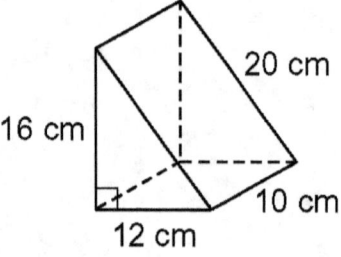 _____

10. 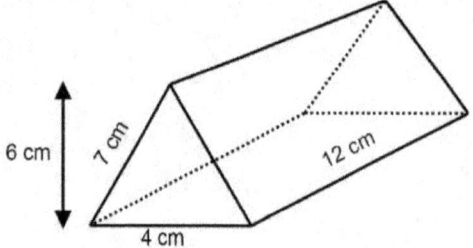 _____

24.4 Nets

Some 3D shapes, like cubes and pyramids, can be opened out and unfolded into a flat shape. The unfolded shape is called the net of the solid.

The flat shape must contain the same shapes as the faces of the 3D shape.

Exercise 24.4
Name the shape that these nets would fold into.

1. 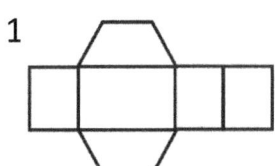 2. 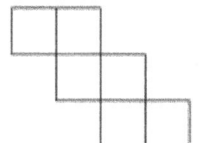 3. 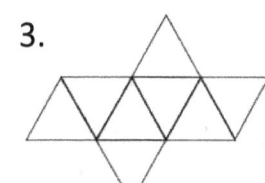 4.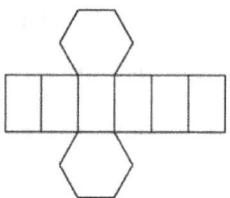

_____   _____   _____   _____

_____   _____   _____   _____

5. 6. 7. 8.

      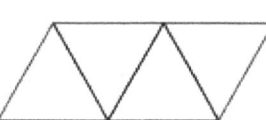   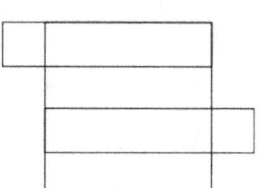

_____   _____   _____   _____

_____   _____   _____   _____

9. 10.

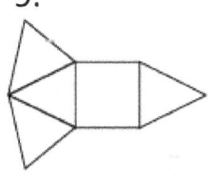   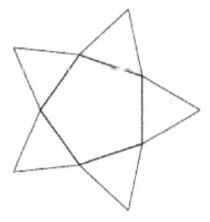

_____   _____

_____   _____

Puzzle Page 4

How do you keep warm in a cold room?

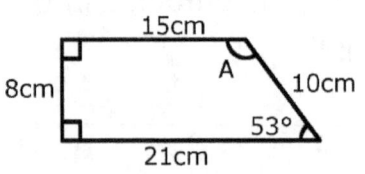 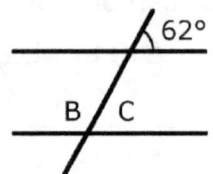 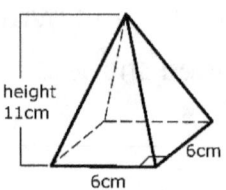 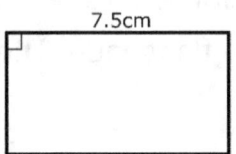

5782 + 364 + 97 − 12 =   6231   b

1. Find the perimeter of the trapezium in cm.
2. Find the area of the trapezium in cm^2.
3. Find the size of angle A.
4. Find the size of angle B.
5. Find the size of angle C.
6. Find the volume of the square-based pyramid in cm^3.
7. Find the surface area of the square-based pyramid in cm^2. (Use area of triangular side = 34cm^2)
8. The rectangle has an area of 30cm^2. Find the length of the missing side.
9. What is the perimeter of the rectangle?
10. If a square has a perimeter of 20cm, what is its area in cm^2?
11. What is the volume of a hexagonal prism, with a base of 8cm^2 and a height of 14cm?

A square park with a side of 25m has a path with a width of 2m around it.

12. What is the outer perimeter of the path?

Work out what letter each of your answers represents below:

4	23	25	54	62	112
o	g	i	a	r	e

6231	116	118	127	132	144	172
b	s	h	n	c	d	t

Now write the letter above the question numbers, below to answer the puzzle.

__ __ __ __ __   __ __   __ __ __   __ __
12 7 1 3 2   10 3   7 4 11

__ __ __ __ __ __   __ __ ' __   9 0
 6 8 5 3 11 5   10 7   12

__ __ __ __ __ __ __   __ __ __ __ __
 2 11 9 5 11 11 12   7 4 11 5 11

Chapter 25: Navigation

25.1 Compass Points

The main compass points are: north, south, east and west. North is always drawn pointing up.

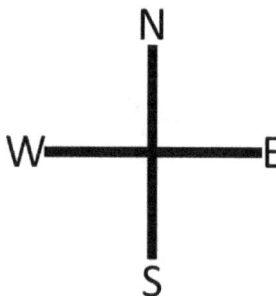

These points can be remembered as only in this direction do you spell a word (we). The order of the compass points can also be remembered by starting at North and moving around clockwise. Any of the following sentences can be used:

- Naughty elephants squirt water.
- Never eat soggy Weetabix.
- Never eat shredded wheat.

Or, you can make up a sentence of your own.

Between these points we have north-east, south-east, south, west and north-west. Notice that when the two directions are put together, north or south always comes first.

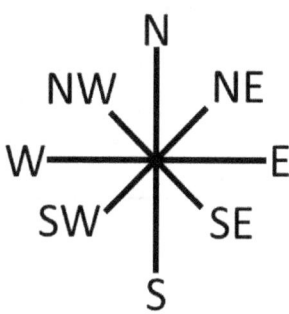

Between each of the main compass points is 90°.

Example: If someone is facing east and turns 270° anticlockwise, what direction are they facing?

Answer: South.

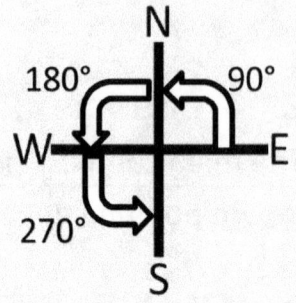

Example: If someone is facing south and turns 45° clockwise, what direction are they facing?

Answer: south-west.

Exercise 25.1

What direction is someone facing if:

1. They face north and turn 90° clockwise? _____

2. They face south and turn 180° anti-clockwise? _____

3. They face west and turn 45° clockwise? _____

4. They face east and turn 270° clockwise? _____

5. They face north and turn 45° anticlockwise? _____

6. They face west and turn 135° clockwise? _____

7. They face south-west and turn 180° clockwise? _____

8. They face north-east and turn 45° clockwise? _____

9. They face south-east and turn 135° anticlockwise? _____

10. They face north-west and turn 270° clockwise? _____

25.2 Directions

Compass points are most commonly used to provide directions.

Example: Walk 3 squares North, 4 squares West and 1 square South. What square do you end up on?

Answer: P

A	B	C	D	E
F	G	H	I	J
K	L	M	N	O
P	Q	R	S	T
U	V	W	X	Y
Z	1	2	3	START

Exercise 25.2

Use the grid below to answer the questions.

A	B	C	D	E	F	G	H	I	J
K	L	M	N	O	P	Q	R	S	T
U	V	W	X	Y	Z	AA	AB	AC	AD
AE	AF	AG	AH	start	AJ	AK	AL	AM	AN
AO	AP	AQ	AR	AS	AT	AU	AV	AW	AX
AY	AZ	BA	BB	BC	BD	BE	BF	BG	BH
BI	BJ	BK	BL	BM	BN	BO	BP	BQ	BR
BS	BT	BU	BV	BW	BX	BY	BZ	CA	CB

1. Walk 2 squares south and 3 east.
2. Walk 3 squares east and 2 north.
3. Walk 4 squares west, 2 squares north and 6 squares east.
4. Walk 3 squares north and 5 squares south.
5. Walk 4 squares south, 3 squares west and 2 squares north.
6. Walk 3 squares east, turn 90° clockwise, walk 2 squares.
7. Walk 2 squares north, turn 270° clockwise, walk 3 squares.
8. Walk 3 squares south-east.
9. Walk 2 squares north-west, turn 135° anticlockwise, walk 2 squares.
10. Walk 3 squares north-east, 5 squares south and 5 squares west.

25.3 Speed

Speed is the distance travelled in a particular time.

Speed can be calculated using the formula:

$$speed = \frac{distance}{time}$$

This can be written as:

$$s = \frac{d}{t}$$

The units of speed can be:

- miles per hour – mph
- kilometres per hour – km/h
- metres per second – m/s

Example: If a cyclist travels 25km in 2 hours. What is their average speed?

$$s = \frac{d}{t}$$
$$= \frac{25}{2}$$
$$= 12.5 \text{ km/h}$$

Example: If a car travels at 30mph for 8 hours, what distance will they have travelled?

$$s = \frac{d}{t}$$
$$30 = \frac{d}{8} \qquad \text{multiply both sides by 8}$$
$$30 \times 8 = \frac{d}{8} \times 8$$
$$d = 240 \text{ miles}$$

Exercise 25.3

1. A person walks 18km in 3 hours. What was their average speed? _____

2. A type of bird called a swift. When migrating a swift can travel 900km in a day. What is its average speed in km/h? _____

3. If an elephant walks 90 miles in 6 hours. What speed is it walking? _____

4. Joel's Ferrari goes a distance of 380 km in a time of 2 hours. What was his average speed in km/h? _____

5. Brett and Brenda are going to a party together. They travel at a speed of 4.2 m/s for 62 seconds. How far away is the party? _____

6. A cheetah can cover 100m in 6 seconds. How fast does it run? _____

7. Harry and George are walking to the supermarket to buy some juice. They walk 1000 metres in 900 seconds. What was their average speed in km/h? _____

8. The Peregrine Falcon can travel 300km/h, in a dive. How far can it travel in one minute? _____

9. Whitney drives 30 km at 45km/h. How long did her journey take in minutes? _____

10. A space ship travels 234000000 kilometres in 15 hours. What was the average speed of the space ship in km/h? _____

25.4 Distance-time graphs

Distance and time can be plotted on a graph, with time on the x-axis.

On a distance time graph:

- A straight line means a constant speed.
- A horizontal line means they are stopped. (distance is staying the same)
- The steeper the line the faster the speed.

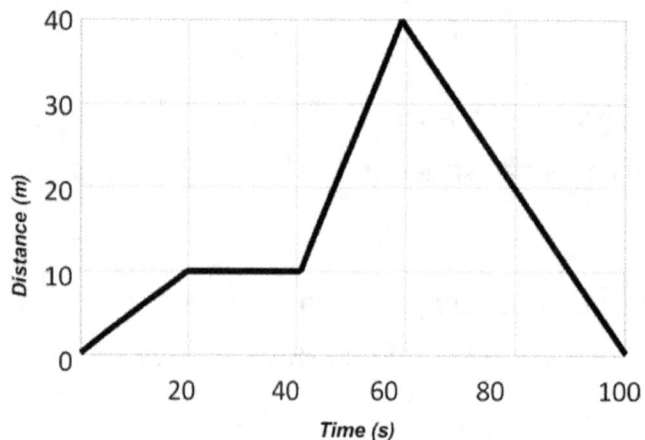

Looking at the above graph. To describe the journey: They travelled for 20s at a constant speed. They then stopped for 20 seconds. They then travelled for 20 seconds at a faster speed than before. They then reversed direction and returned to the start in 40 seconds.

Speed can be calculated from a distance time graph.

Example: Calculate the speed at 5 seconds, in the graph above.
5 seconds is on the line from 0 to 20 seconds.

$$s = \frac{d}{t}$$

$$= \frac{10}{20}$$

$$= 0.5 \text{ m/s}.$$

Exercise 25.4

Use the graph to answer the following questions.

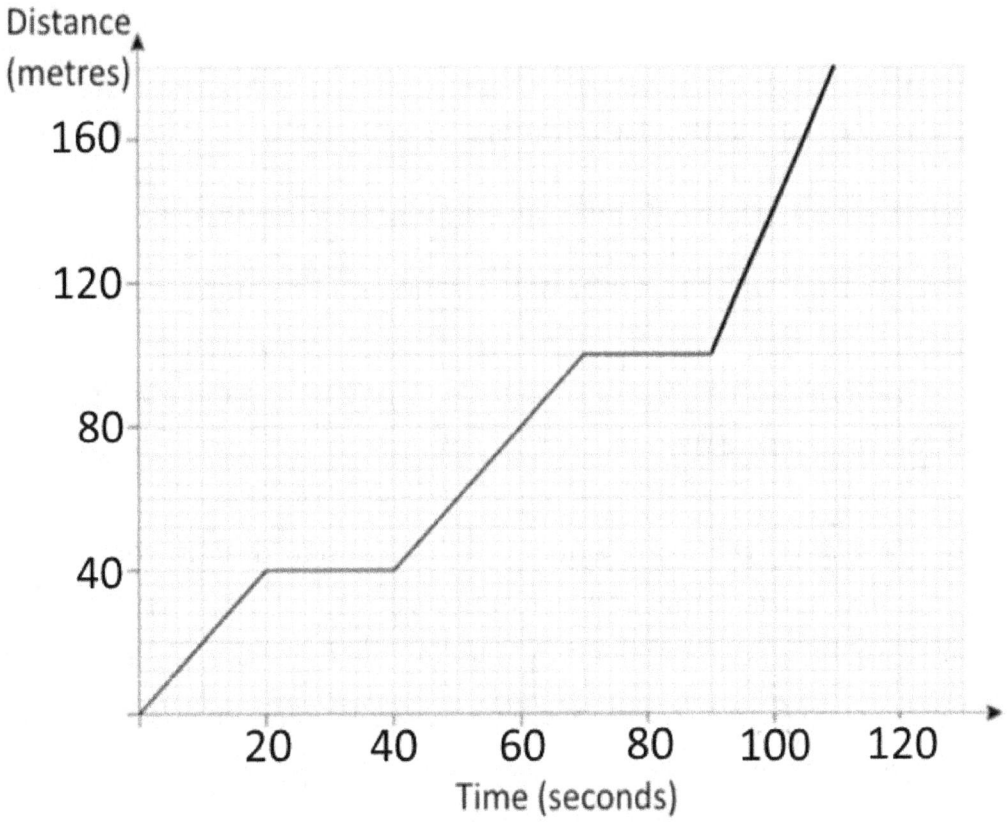

1. How long did it take to travel 100m? _____

2. How far did they travel in the first minute? _____

3. What is happening at 20s? _____

4. How far have they travelled in 100s? _____

5. What speed are they travelling at 10s? _____

6. What speed are they travelling at 45s? _____

7. What is the speed at 30s? _____

8. Without any calculations compare the speed at the end of the trip, to the beginning. _____

9. How can you tell? _____

10. What speed are they traveling at 100s? (after their second rest) _____

Chapter 26: Ratios

26.1 Ratios

A ratio is a way of comparing two or more values.

The different values are separated by a colon (:) or the word "to".

Ratios have an order and the order of the ratio and the numbers must match.

So if there are: four apples, three oranges and one banana in a fruit bowl. We can write the ratio as 4:3:1.

Ratios are *always* written using whole numbers and in their lowest terms and can be simplified in the same way as fractions.
So the ratio 30:20 can be simplified by dividing each number by 10, giving the ratio 3:2. The ratio $\frac{3}{5}$: 2, can be simplified by multiplying both sides by 5, giving the ratio 3:10.

Ratios can be converted into fractions, by putting the part over the total (the sum of the parts). We can then convert the fraction into a percentage by multiplying by 100.

Example:

In a class there are 22 children who prefer dogs to cats and 16 who prefer cats. What is the ratio and the fraction of children who like dogs and cats best?

The ratio of those who prefer dogs to those who prefer cats: 22:16
But, ratios are always in their lowest terms, so divide each side by 2.

$$22:16$$
$$\div 2 \ \div 2$$
$$11:8$$

> To convert to a fraction, add the two sides to get the denominator: 11+8 = 19.
>
> Now, put the number in the ratio over 19 (the denominator).
>
> So fraction of children who prefer dogs is $\frac{11}{19}$ and those who prefer cats is $\frac{8}{19}$.

Exercise 26.1

Write the following ratios (remember ratios are always written in their lowest terms).

1. Last week, Desmond took ham sandwiches to school on two days and cheese sandwiches on the other three days. What is the ratio of ham to cheese sandwiches? _____

2. A pet store has 8 gerbils, 3 guinea pigs and 5 chinchillas for sale. What is the ratio of gerbils to chinchillas to guinea pigs? _____

3. If after a party there are $3\frac{1}{2}$ apple pies left and $1\frac{3}{4}$ berry pies left, what is the ratio of the remaining apple pies to berry pies? _____

4. If a daily newspaper contains 35 pages but the weekend edition contains 80 pages, what is the ratio of the number of pages during the week compared to the weekend edition? _____

5. The ratio of children to adults at a function is 1:2. What is the fraction of children at the party? _____

6. The ratio of flour to milk to eggs in a cake is 9:3:1. What fraction of the cake mixture is milk? _____

7. Yesterday, the ratio of the time Elise was sleeping to the time she was awake is 3 : 5. What fraction of the day did she spend sleeping? _____

8. On Monday Martha made flapjacks and brownies in the ratio of 5:7. What fraction of the baking was flapjacks? _____

9. Cement can be made by mixing sand and cement in the ratio 3:1. What is the percentage of sand in the mixture? _____

10. The ratio of wet days to fine days on a camping trip was 3:2. What was the percentage of fine days? _____

26.2 Ratio Problems

Ratios are in direct proportion to the amounts. So whatever we multiply or divide one part by, to the other parts we do *exactly the same*.

To work out the multiplier, we can work out the total number of parts in the ratio and compare to the total.

Example

The label of a squash bottle says to add 1 part squash to 4 parts water (ie. a ratio of 1:4). If there is 60mL of squash left in the bottle, how much water should be added?

Answer

Ratio	1 : 4	We know that the total for squash is 60.
	x60 x60	To go from the 1 in the ratio to 60 we need to multiply by 60.
	60: **240**	So to the water we do exactly the same.

Therefore 240mL should be added.

Example

A recipe mixes flour and butter in the ratio 7:5. If we make 360g in total, how much butter do we need?

Ratio 7:5   Total of parts 12

x30 x30   x30

210:150   360

Therefore 150g of butter is needed.

Exercise 26.2

1. In a secondary school the ratio of boys to girls is 11:12. If there are 720 girls in the school, how many boys are there? _____

2. A recipe for a crumble mixes butter, flour and sugar in the ratio 3:4:2. If Marion want to make 270g of mixture, how much butter does she need? _____

3. The total number of marks for a test is 40. The marks are divided between Section A and Section B in the ratio 4 : 1. How many marks are there for Section B? _____

4. Mark and John work 4 hours and 7 hours respectably. The total pay was $99. If they each received the same hourly pay, how much pay did John receive? _____

5. A piece of wood is of length 45 cm. The length is divided into the ratio 7 : 2. What is the length of each part? _____ & _____

6. Joel, Dominic and Bill share 36 sweets in the ratio 2 : 3 : 4.
   Work out the number of sweets that Dominic receives.

7. Hamza is going to make some concrete mix. He needs to mix cement powder, sand and gravel in the ratio 1 : 3 : 5 by weight. Hamza wants to make 180 kg of concrete mix, how much cement powder does he need?

8. A shop sells CDs and DVDs. In one week the number of CDs sold and the number of DVDs sold were in the ratio 3:5. The total number of CDs and DVDs sold in the week was 728. How many CDs were sold that week?

9. For a drink, Jenomi mixes orange cordial and lemonade in the ratio 1 : 5. If she wants to make 900mL of this drink, how much lemonade should she use?

10. Colin has only 5p coins and 10p coins. The ratio of the number of 5p coins to the number of 10p coins is 2 : 3. Work out the ratio of the total value of the 5p coins : the total value of the 10p coins

26.3 Maps and Scales

One common use of ratios is in maps and scale drawings.

In these questions you do not need to convert the measurements to the same units as you do with other ratio questions.

If the scale on a map is 3cm:8km then every 3cm on the map will represent 8km.

Example

The scale on a map is 3cm:8km. If the distance travelled on the map measures 27cm, how far have they travelled?

    Scale   3cm:8km
             x9   x9

        27cm: 72km

Therefore they have travelled 72km.

Example

A plan is drawn for a chicken coop. The scale is 1cm:2ft.
What will the length of the run measure on the plan?

   scale 1cm:2ft

       x4   x4

   4cm:8ft

So the length on the plan is 4cm.

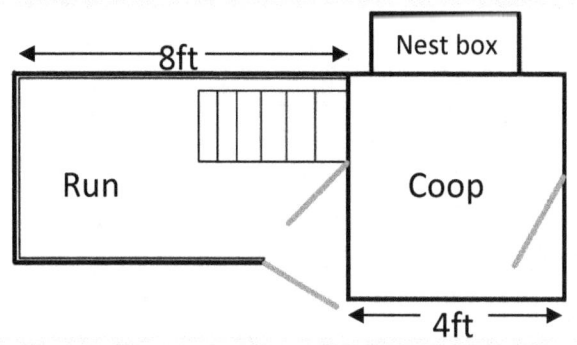

Exercise 26.3

1. A map has a scale 1 cm : 5 km. The distance between Liverpool and Glasgow is 56cm on the map. What distance does this represent? _____

2. A map has a scale of 3cm:10mi. If Birmingham to Nottingham is 50 miles, what distance will it be on the map? _____

3. If the human is 6ft and drawn to a height of 1.8cm. What is the scale? _____

4. The allosaurus is drawn to the same scale. If the dinosaur is drawn with a length of 12cm, how long is the dinosaur? _____

5. Su draws a picture of the Eiffel tower with a scale of 2cm:75m. If the Eiffel tower is 300m, how tall is Su's drawing? _____

6. Ryan is building this chest of drawers. He needs to know how long the pieces of wood on the sides are. If the diagram measures 16cm, how tall is the chest? _____

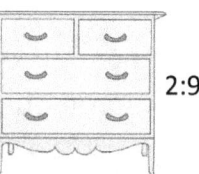

2:9

7. scale → 1 in. : 3 ft

The bicycle shown is drawn with a length of 2". What is the length of the bicycle? _____

8. A map of the UK has a scale of 1:800 000. What distance is represented by 1cm on the map? _____

9. A school has a charity sports event and all the events are drawn on a diagram of the field. If the diagram has a width of 6cm and is drawn with a scale of 3cm:80ft, what is the width of the field? _____

10. Ben draws a scale drawing of his room. His wardrobe has a width of 130cm and is 6.5cm on his drawing. What is the scale of his diagram? _____

26.4 Hidden Ratios

Some questions do not mention ratio or scale but the easiest way to do them is by creating a ratio.

To do this:

- Make the smallest quantity one.
- Multiply, as necessary, so that all numbers are whole.

Example

Amy is 5 times younger than her mother, who is half the age of Amy's great grandmother. Their total ages are 160, how old is Amy's great grandmother?

First – work out the ratio. Amy is the youngest, so she will be one. Her mother is five times older, so will be 5. Her great-grandmother is twice as old again so will be 10.

$$\begin{array}{cc} \text{This gives the ratio} \ 1:5:10 & \text{Sum of the parts is 16} \\ \text{x10 x10 x10} & \text{x10} \downarrow \\ 10:50:100 & 160 \end{array}$$

So Amy's Great Grandmother is 100 years old.

Example

In an hour, Katrina spends twice as long playing the piano as she does eating. How long does she spend playing the piano?

$$\begin{array}{cc} \text{The ratio of piano to eating is} \ 2:1 & \text{total = 3} \\ \text{x20 x20} & \text{x20} \downarrow \\ 40:20 & \text{60 minutes} \end{array}$$

Katrina spends 40 minutes practicing the piano.

Exercise 26.4

1. Sam has a brother Max. Max is six times older than Sam. Their Dad is four times Max's age. Their ages add up to 62. How old is Max? _____

2. Jamie is sorting out his Lego™ collection. He has three times more red bricks than white bricks. If he has a total of 160 bricks, how many white bricks does he have? _____

3. He has $\frac{1}{4}$ the number of roof tiles as he has bricks. How many roof tiles does he have? _____

4. On a coach trip Penny stops at Weymouth. From Weymouth it is ¼ the distance to Poole or ½ the distance to Southampton as it is to Brighton. If she decides to travel on to Brighton the next day, how far will she travel? (Total distance from Weymouth to each destination is 350km) _____

5. In a raffle, the value of first prize is five times the value of second, which is three times the value of third prize. If the total value of the prize pool is $1900, what is the value of first prize? _____

6. Tim is nineteen times older than Jay and their combined age is 100. How old is Tim? _____

7. In an hour exam, Graeme spends twice as long on question 2 and three times as long on question 3 as he does on question 1. How long does he spend on question 1? _____

8. Rupa practices the piano for two hours. If she spends half the time on her scales as she spends on her exercises and one and a half times more on her exam pieces, how long does she spend on her scales? _____

9. Tigger, Pooh and Piglet between them walk (or bounce) a combined distance of 5 miles, in the Hundred Acre Wood, to visit Christopher Robin. Tigger bounces three times further than Pooh and Piglet who walk together chatting. How far does Pooh walk? _____

10. Jasmine, Aladdin and Jafar enter a read-a-thon to raise money for charity. They read a total of 135 books. Jafar reads ten times more books than Jasmine, who reads four times more than Aladdin. How many books does Aladdin read? _____

Chapter 27: Probability

27.1 What is Probability

Probability is how likely something is to happen. Probability can range from impossible to certain. This can be expressed in words or in numbers.

certain --- highly likely --- likely --- even chance --- unlikely --- highly unlikely --- impossible

The word possible can also be used to express the probabilities between highly likely and highly unlikely.

As well as words, we can use numbers between zero and one.

- 0 is impossible
- 0.5 is even chance
- 1 is certain

Before calculating probability it is often useful to know if the situation is fair or not. A situation is fair if all outcomes are equally likely. For example, normal dice are fair because there is an equal chance of throwing a 1, 2, 3, 4, 5 or 6.

Exercise 27.1

To answer questions 1 – 4 use the words: certain, likely, equal chance, unlikely and impossible.

What is the probability that:

1. The sun will rise tomorrow? _____

2. I flip a coin and it lands heads up? _____

3. A letter chosen at random from the alphabet is a consonant? _____

4. Two dice are thrown and both land on a six? _____

Answer questions 5 – 7 as a number between 0 and 1 (inclusive).

5. Tomorrow you will fly to Mars?
6. After Tuesday is Wednesday?
7. A new baby will be a boy?

8. The probability that a bus will be late is 0.1. Which of these statements is the most reasonable?
    a. The bus is unlikely to be late.
    b. The bus is likely to be late.
    c. The bus is certain to be late.

State whether the following situations are fair or unfair.

9. A spinner with six even sides.
10. A die that is weighted so it always lands on a six.

27.2 Calculating Probability

If all outcomes are equally likely, then the probability of an outcome can be determined by the formula:

$$Probability\ (event) = \frac{Number\ of\ ways\ it\ can\ occur}{Total\ number\ of\ outcomes}$$

Example

If a die is thrown, what is the probability it will be a five or six?

$$P(5\ or\ 6) = \frac{5\ or\ 6}{all\ numbers}$$

$$= \frac{2}{6}$$

$$= \frac{1}{3}$$

A standard pack of playing cards has 52 cards in 4 suits. Two of the suits are red (hearts and diamonds) and two of the suits are black (clubs and spades). Each suit has 13 cards: ace, 2, 3, 4, 5, 6, 7, 8, 9, jack, queen and king. The card marked as one is called an ace. Jack, queen and king are each worth 10. Some packs also have two jokers, but only include these if they are specifically mentioned in the question.

The probability of an even not happening is one minus the even happening.
P(even not happening) = 1 − P(even happening)

Example

What is the probability that if a dice is thrown the number is not a six.

P(not a six) = 1 − P(6)

$$= 1 - \frac{1}{6}$$

$$= \frac{6}{6} - \frac{1}{6}$$

$$= \frac{5}{6}$$

Exercise 27.2

Calculate the probability for the following questions. Give your answers as fractions.

1. Throwing a die and getting an even number? _____

2. A raffle has 150 tickets. If Anne buys 30 tickets what is the probability that Anne will win the raffle? _____

3. A school has four year seven tutor groups: 7A, 7B, 7R and 7S. What is the chance that Willem will be put in 7A? _____

4. If Rose takes a card at random from a standard pack of cards, what is the chance that she will pick a jack, queen or king of any suit? _____

5. If Paul takes a card from a standard pack, what is the chance that he will take a red ace? _____

A bag contains 24 balls: 9 red, 4 green, 3 black and 8 white. If I take a ball at random, what is the chance that I will take a:

6. Black ball? _____

7. Red ball? _____

8. White ball? _____

9. Green ball? _____

10. A ball that is not white? _____

27.3 Outcomes of Combined Events

When two events are considered together, it is important to determine if:

- The first outcome can occur again.
- If the events are independent – that is one outcome does not affect the other outcome.

For example: I have a bag containing 6 balls – 2 red, 2 blue, 1 green and 1 yellow. If I take out one ball and then return it and then take out a second ball, the colour of my first ball will not affect the colour of my second ball. So the events are independent.

However, if I take out one ball, keep it and then take out a second ball, the colour of the first ball will affect the colour of the second ball. For example, if the first ball is yellow, my second ball cannot be yellow. So the events are dependent. If it is a red ball the chance of bringing out a second red ball is decreased, while the chance of any other colour is increased.

It is really important, when listing possible outcomes **to do so logically**. Otherwise, a lot of time can be wasted trying to work out which outcome was missed.

Example

If two balls are drawn out of a bag containing red (R), blue (B), and green balls (G), then the first ball can be red and the second can be red, blue or green, then the first ball can be blue and so on, giving

RR, RB, RG, BR, BB, BG, GR, GB, GG

If people are chosen for sports games, it's important to remember that they cannot play against themselves.

Outcomes can also be put into a table. Here is a table of the outcome of two dice being thrown.

	6	7	8	9	10	11	12
	5	6	7	8	9	10	11
	4	5	6	7	8	9	10
2nd die	3	4	5	6	7	8	9
	2	3	4	5	6	7	8
	1	2	3	4	5	6	7
		1	2	3	4	5	6

1st die

If I have one event followed by another, then the number of combinations is:

Number of event 1 outcomes x number of event 2 outcomes

Example:

If for a meal there are three choices of entrée, four choices of main and four choices of dessert, how many possible meal combinations are there?

Number of combinations = no of entrée x no of main x no of dessert

= 3 x 4 x 4 = 48

Exercise 27.3

1. Ivan tosses two coins. What are the possible results?

   _____, _____, _____

2. Andrew (A), Ben (B), Carl (C) and David (D) play tennis. If each player plays every other player, what are the games that are played? (Write AB for Andrew playing Ben).

   _____

A bag contains 4 balls: 2 blue (B), 1 red (R) and 1 green (G)

3. What are the possible outcomes if Kate pulls out one ball, puts it back in the bag and pulls out a second ball? (Consider RB and BR to be different)

   _____

4. What are the possible outcomes if Kate removes one ball, keeps it and removes a second ball? _____

   _____

5. What are the possible outcomes if Kate removes two balls at the same time (order doesn't matter)? _____

6. Séamus, Philip, Judy and Tracey belong to a chess club. If each player plays every other player once, how many games are played? _____

7. A conference for breakfast has fruit, porridge or toast and tea, coffee, hot chocolate and orange juice. If someone has one food item and one drink, how many combinations are there to choose from? _____

8. James has 70 books and 8 toys. If he takes one book and one toy away on holidays, how many possible combinations are there? _____

9. If there are five different types of tea and three different types of biscuit, how many different combinations or tea and biscuit are there? _____

10. On a camp there is a choice of five outside activities before lunch and four quieter indoor activities after lunch. How many different combinations of activities are available? _____

27.4 Probability of Combined Events

To calculate the probability of one event happening followed by a second event, multiply the two probabilities together.

Example

What is the probability of rolling a dice twice and rolling a 6 followed by a 4?

P(6 then 4) = P(6) x P(4)

$$= \frac{1}{6} \times \frac{1}{6}$$

$$= \frac{1}{36}$$

If the outcomes can happen in either order, then add up the two (or more) possibilities.

Example

What is the probability of rolling two dice and throwing a 6 and a 4?

P(6 & 4) = P(6 then 4) + P(4 then 6)

$$= \frac{1}{36} + \frac{1}{36}$$

$$= \frac{2}{36}$$

$$= \frac{1}{18}$$

If you can work out the possible outcomes, then use the formula for a single event.

> Example
>
> What is the probability of rolling two dice and throwing a total of 10?
>
> Using the table in the last section there are 3 possibilities that give a total of 10.
>
> $$P(10) = \frac{nunber\ of\ favourable\ outcomes}{total\ number\ of\ outcomes}$$
>
> $$= \frac{3}{36}$$
>
> $$= \frac{1}{12}$$

Exercise 27.4

Use the table of the two dice in the last section, as appropriate.

1. If I throw a coin twice, what is the probability that I will get two heads?

2. If I throw two dice, what is the probability that I will get two sixes?

3. What is the probability that I will not get two sixes?

4. If I throw two dice, what is the probability that I will get a total of eight?

5. If I throw a coin twice, what is the probability that I will get one head and one tail?

6. If Ray, Moye, Stan, Terrence, and Huan play tennis. They draw names out of a hat to see who will play. What is the probability that the first game will be Stan and Terrence?

Kareem has 2 bags of sweets.
Bag A contains 6 brown sweets and 5 red sweets.
Bag B contains 2 brown sweets and 4 green sweets.

7. Kareem takes one sweet at random from each bag.
   What is the probability that he takes two brown sweets? _____

8. If he takes two sweets from bag B, what is the probability that
   he will choose both brown sweets? _____

5 white socks and 3 black socks are in a drawer.
Priya takes out two socks at random.

9. Work out the probability that Priya takes out two white socks? _____

10. Work out the probability that Priya takes out two socks of the same
    colour. _____

Chapter 28: Tables and Graphs

28.1 Two Way Tables

Data can be presented in tables and graphs. Tables consist of data in columns and rows. Any units are normally in the column headings. To find a piece of data look at where the appropriate row and column intersect.

Calculations can also be made to work out totals and missing numbers.

Example

Year 5 students were given a choice of ice-cream flavours. The flavours chosen are shown in the table below:

	Boys	Girls
Chocolate	12	17
Vanilla	5	3
Strawberry		9
Caramel	3	1
Banana	2	0
Total	30	30

How many girls ordered a vanilla? Simply look down the girls' column and across the vanilla row. The answer is 3.

How many boys ordered strawberry? The number's missing. We know that the total number of boys was 30. Add up the other flavours – 12+5+3+2 = 22. Take away the other flavours from the total. 30-22 = 8. 8 Boys ordered strawberry.

How many children ordered a chocolate flavoured ice-cream?
Add up the boys and girls in the row labelled chocolate. 12 + 17 = 29.
29 children ordered chocolate.

What fraction of children ordered vanilla?
Eight children out of a total of sixty ordered vanilla.

$$\frac{8}{60} = \frac{2}{15}$$

Exercise 28.1

A survey asked what pet Year 6 children had at home. The findings are shown in the table below.

Pet	Class A	Class B	Class C
Dog	6	7	5
Cat	7	5	4
Rabbit	4	3	4
Guinea pig	2	1	0
Chinchilla	1	2	3
More than one	2	1	2
None	3	4	5

1. In which class do more people have a cat than a dog? _____

2. How many people in year 6 had a chinchilla? _____

3. In class A how many more students had a dog than a rabbit? _____

4. How many more students had a dog than a cat? _____

5. There are six times more dogs than which animal? _____

120 children went on a schools activity day. In the morning they did one activity which was bowling, going to the cinema or ice-skating. The number of students in the activities is shown in the table below.

	Boys	Girls	Totals
Bowling		28	
Cinema		20	36
Skating	15		
Total		66	120

6. How many boys went bowling? _____

7. What fraction of the children who went to the cinema were boys? _____

8. How many children went skating? _____

In a class activity, students had to put a number of 2D shapes, coloured either white or black, into a table. Umer started and his table is shown below. Unfortunately he didn't have time to finish.

	Black	White	Total
3-sided		4	7
4-sided	2		
5-sided		0	2
6-sided	3		5
Total		10	

9. How many white hexagons are there? _____

10. What is the probability that a shape is a white triangle? _____

28.2 Distance Tables

A distance table shows the distances between towns in either miles or kilometres. You simply find the square where the column for one town meets the row for the other town.

Example

The table below shows the distances between classrooms of a school in metres.

Maths			
14	Science		
22	12	English	
40	34	28	History

How many metres between Maths and English? Look for the row where the Maths column meets the English row. The arrows show this. The distance is 22m.

If Troy is in Science. His next lesson is in History followed by English. How far does he need to walk? You need to look for the value for Science to History and the value for History to English and then add them. 34 + 28 = 62m

Exercise 28.2

The table below gives distance in miles between some towns.

Birmingham								
13	Walsall							
17	7	Wolverhampton						
66	75	78	Buckingham					
80	85	98	27	Oxford				
85	91	106	17	25	Aylesbury			
103	109	123	50	27	40	Reading		
109	115	130	60	40	28	26	Slough	
135	140	147	86	65	60	34	26	Guildford

What is the distance between the following towns?

1. Slough and Reading? _____

2. Aylesbury and Wolverhampton? _____

3. Wolverhampton and Birmingham? _____

4. Guildford and Walsall? _____

5. Oxford and Slough? _____

6. Which two towns are closest together?
   _____ and _____

7. From his home in Slough Mr Singh travels to Aylesbury for a meeting and then returns home. How far does he travel? _____

8. Mr Phillips travels from Slough to Oxford via Aylesbury. How far does he travel? _____

9. Elaine lives in Oxford and works in Slough. If she works five days per week, how far does she travel each week to and from work? _____

10. How far is it to travel from Birmingham to Guildford, stopping in Oxford on the way? _____

28.3 Bar Charts

Bar charts are used to display data where one axis is numerical and one axis is not. Axes is the plural of axis.

To determine the number represented by a bar, look across to where the height of the bar would intersect the y-axis. Sometimes, you may need to do some calculations such as subtract the value of one bar from another.

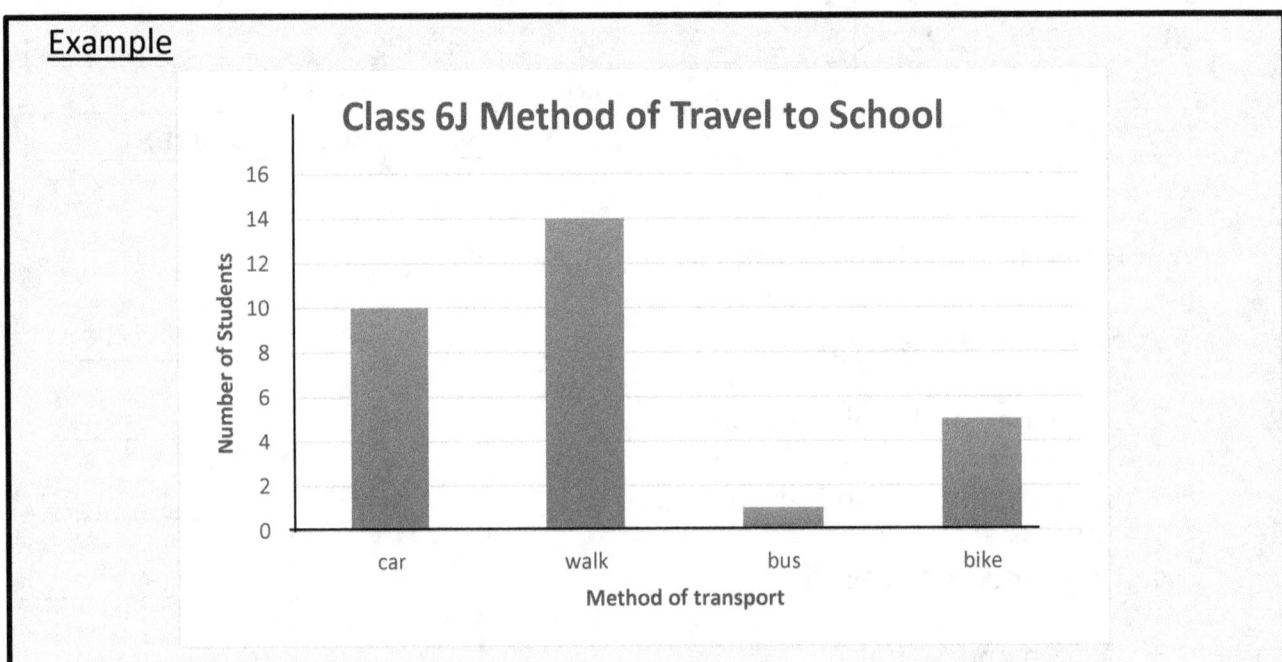

How many students catch a bus to school? Looking across from the top of the bus to the y-axis, the bar is half way between 0 and 2, therefore the number of students that catch a bus to school is 1.

How many more students walk than come by car? 14 students walk and 10 go by car. We need to subtract the number of students who come by car (10) from the number of students that walk (14). Therefore the answer is 4.

Bar charts always have a space between bars. Histograms where both sets of data are numbers, one set in intervals are drawn with no space between bars.

Exercise 28.3

Use the bar chart, showing the favourite fruit of 90 Students in year 5 at Achieve Primary School, to answer the questions.

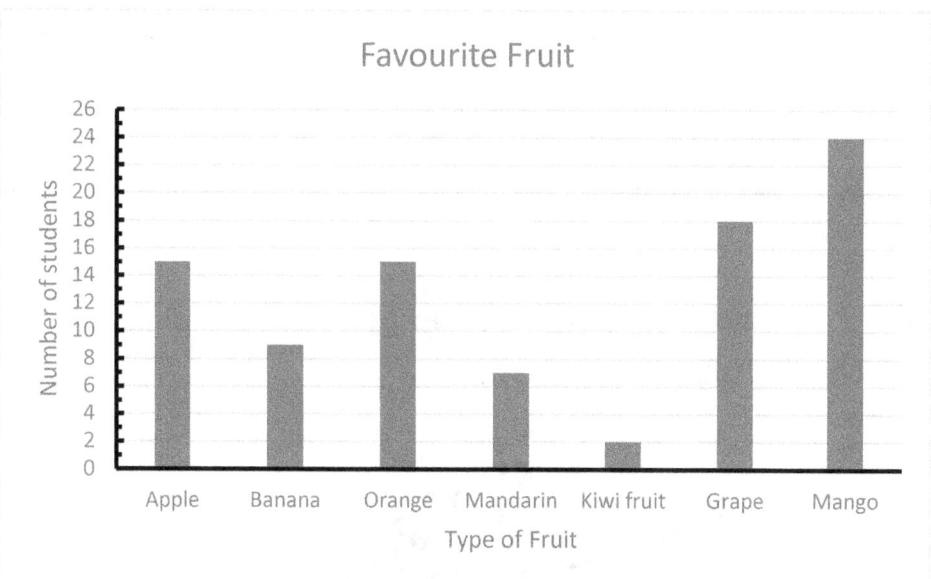

1. How many students like mango best? _____

2. Which is the least popular fruit? _____

3. Which two fruits are equally popular? _____ and _____

4. How many more students prefer oranges to mandarins? _____

5. How many fewer students like kiwi fruit than apples? _____

6. What fraction of students like apples best? _____

7. What fraction of students like bananas or grapes best? _____

8. How many students like mango or grapes best? _____

9. What percentage of students like grapes best? _____

10. What percentage of students like grapes or mango best? (Give your answer to the nearest whole number). _____

28.4 Pie Charts

Pie charts are circles that are cut into slices to show the relative size of different pieces of data. The area of each sector is proportional to the quantity it represents.

Example

Use the pie chart below to answer the questions. 20 children surveyed liked watching TV as their main leisure activity.

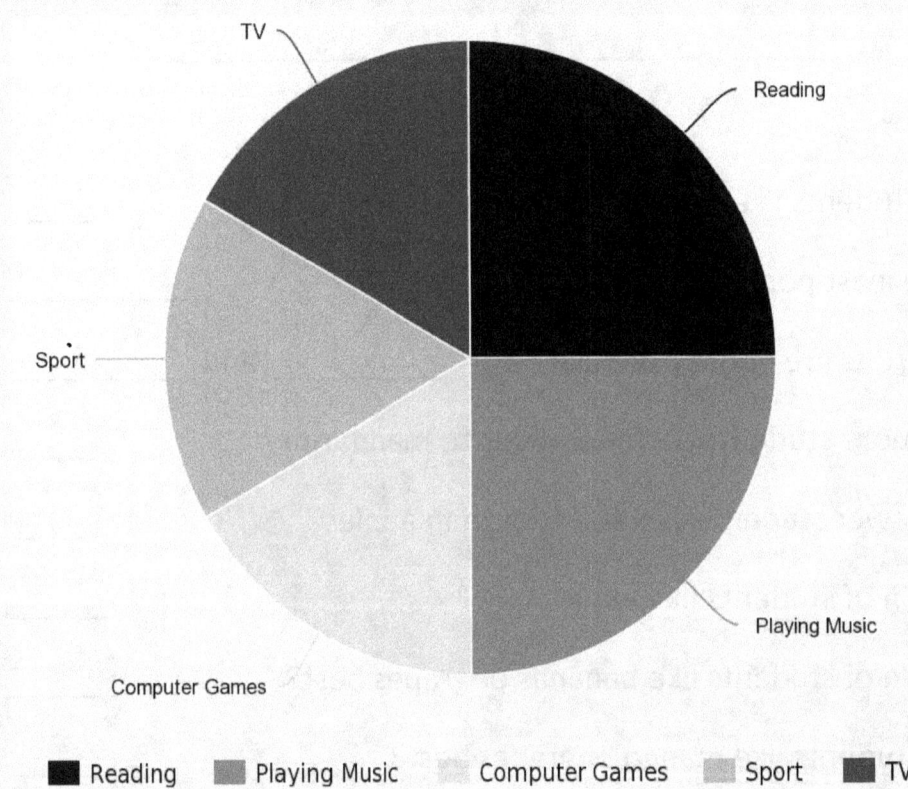

What fraction of children liked playing Sport? Reading and playing music take up half the circle. The other three activities take up an equal space. So the fraction that like sport is $\frac{1}{6}$.

How many students chose reading as their favourite leisure activity?

TV is $\frac{1}{6}$ of the circle and represents 20 children.

If $\frac{1}{6}$ is 20

$\frac{6}{6}$ is 20 x 6 = 120

Reading is $\frac{1}{4}$ of the circle.

(We now know that the whole circle represents 120 children.)

$\frac{1}{4}$ of 120 = 30

Therefore 30 children chose reading as their favourite activity.

Exercise 28.4

The pie chart below shows the number of trees in a small wooded area, which has 360 beech trees as well as a number of oak, ash, pine and yew trees.

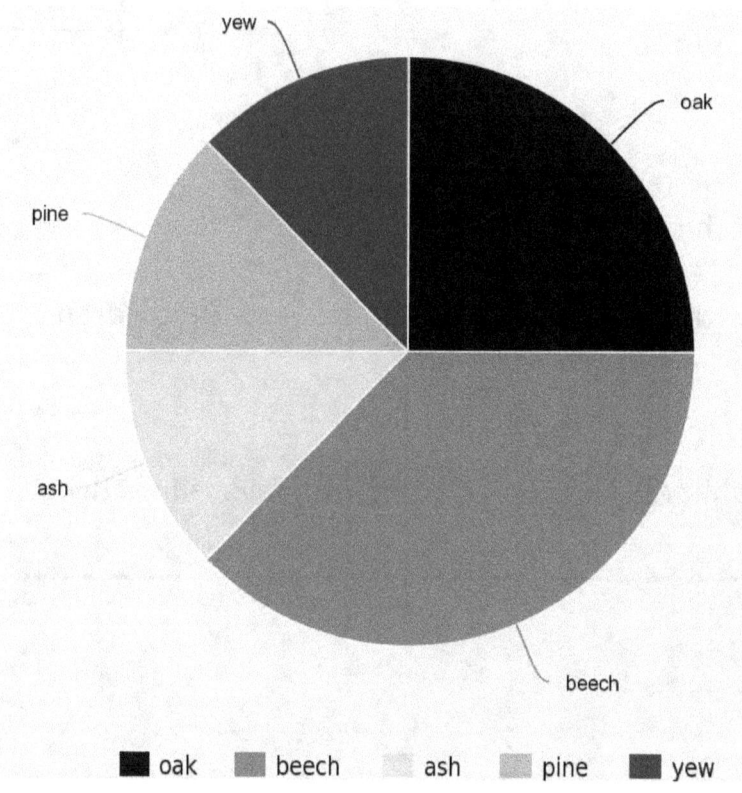

1. What fraction of the trees in the area are ash trees? _____

2. How many oak trees are there? _____

3. How many pine trees are there? _____

4. How many trees are there altogether? _____

5. What percentage of the trees are beech trees? _____

The chart below shows how Elsie spends a typical school day.

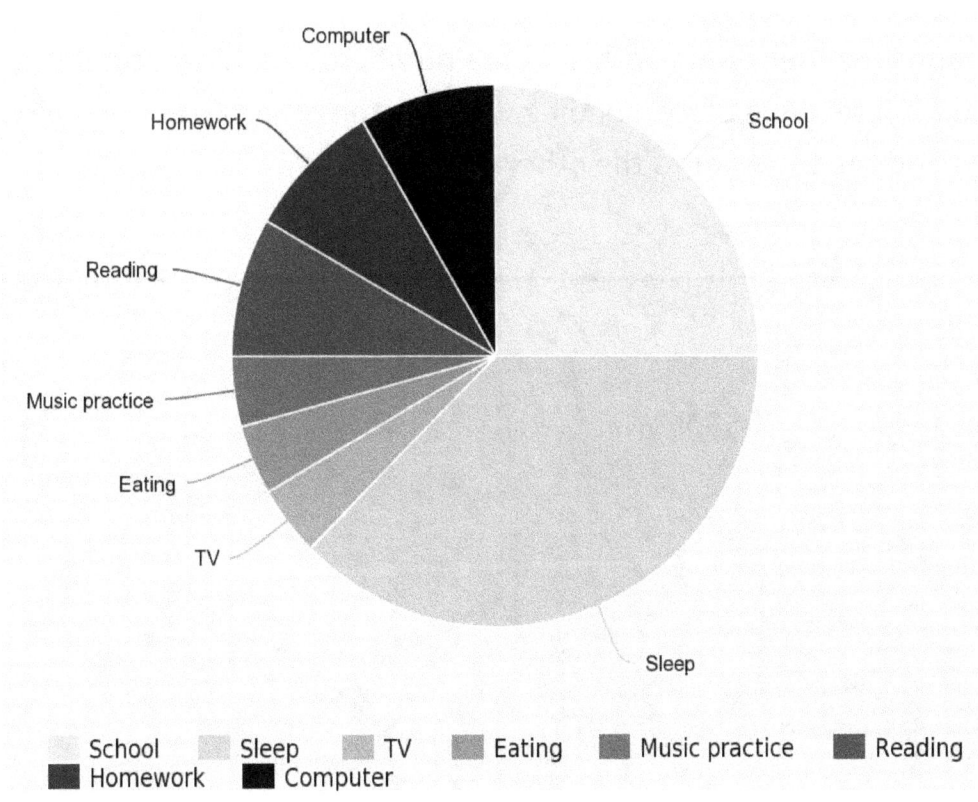

How many hours does Elsie spend

6. At school? _____

7. Reading? _____

8. Sleeping? _____

9. Watching TV? _____

10. Doing homework or practicing music? _____

Chapter 29: Graphs and more

29.1 Line graphs

Line graphs are used when both axes are numbers. For a particular value on one axis, you can read the corresponding value on the other axis by going straight to the line and then straight to the other axis.

Example

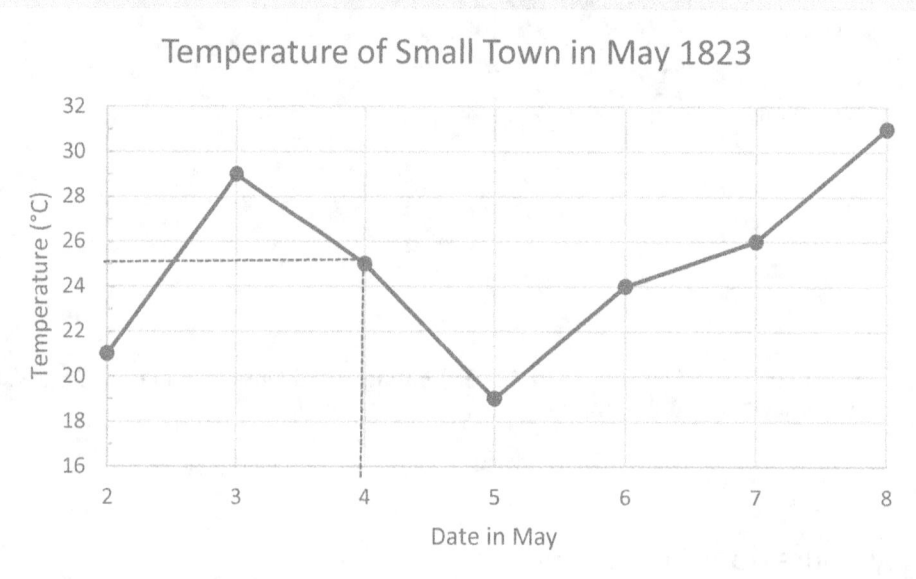

What was the temperature on the fourth of May?

Find the fourth of May (4 on the horizontal axis) and go up until you hit the line, as shown be the dotted line above). Then go across from where you hit the line to the y-axis. It is half way between the 24 and 26. So the temperature on the fourth of May was 25°C.

Exercise 29.1

The chart below shows the height of Ruth from birth to her second birthday.

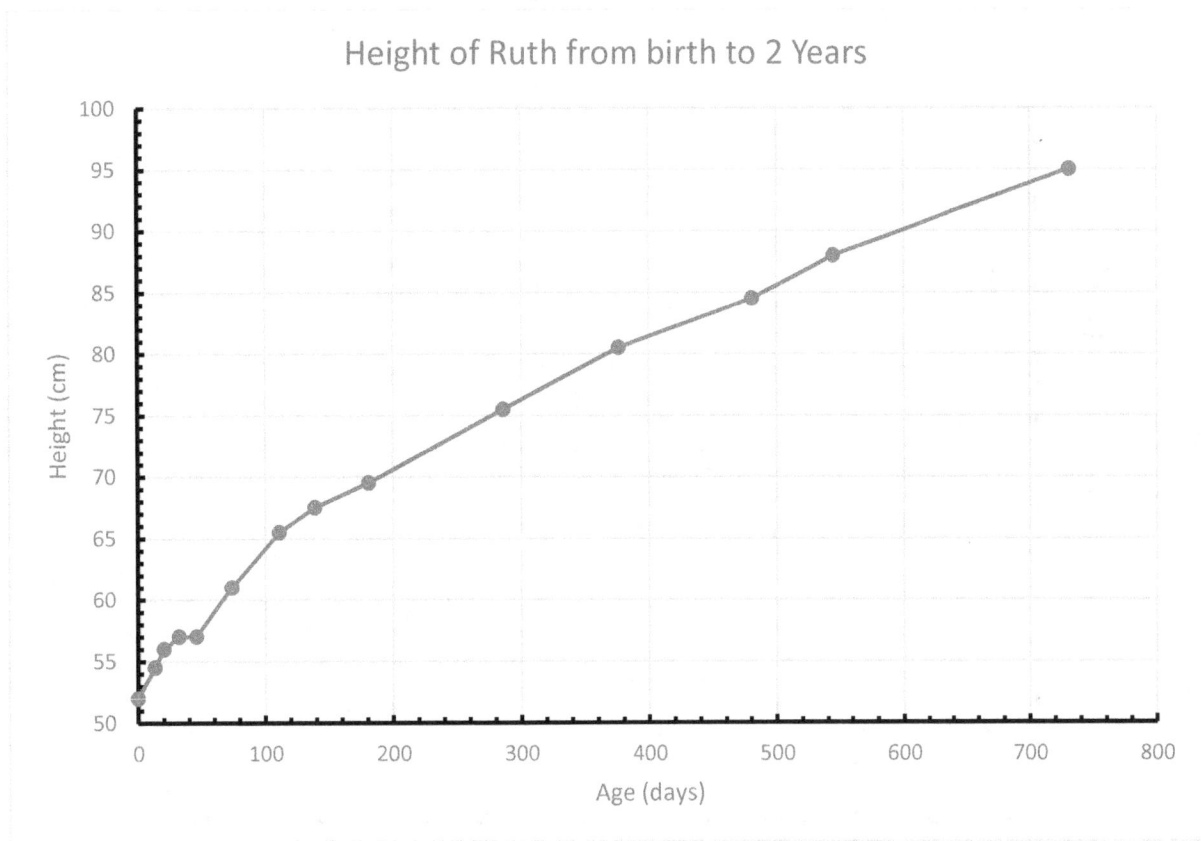

1. What length was Ruth at birth? _____

2. At what age did Ruth reach 75cm? _____

3. What height was Ruth at 500 days? _____

4. What height was Ruth at 1 year? _____

5. The average height of a girl at 2 years is 86cm. How much taller was Ruth than the average 2 year old? _____

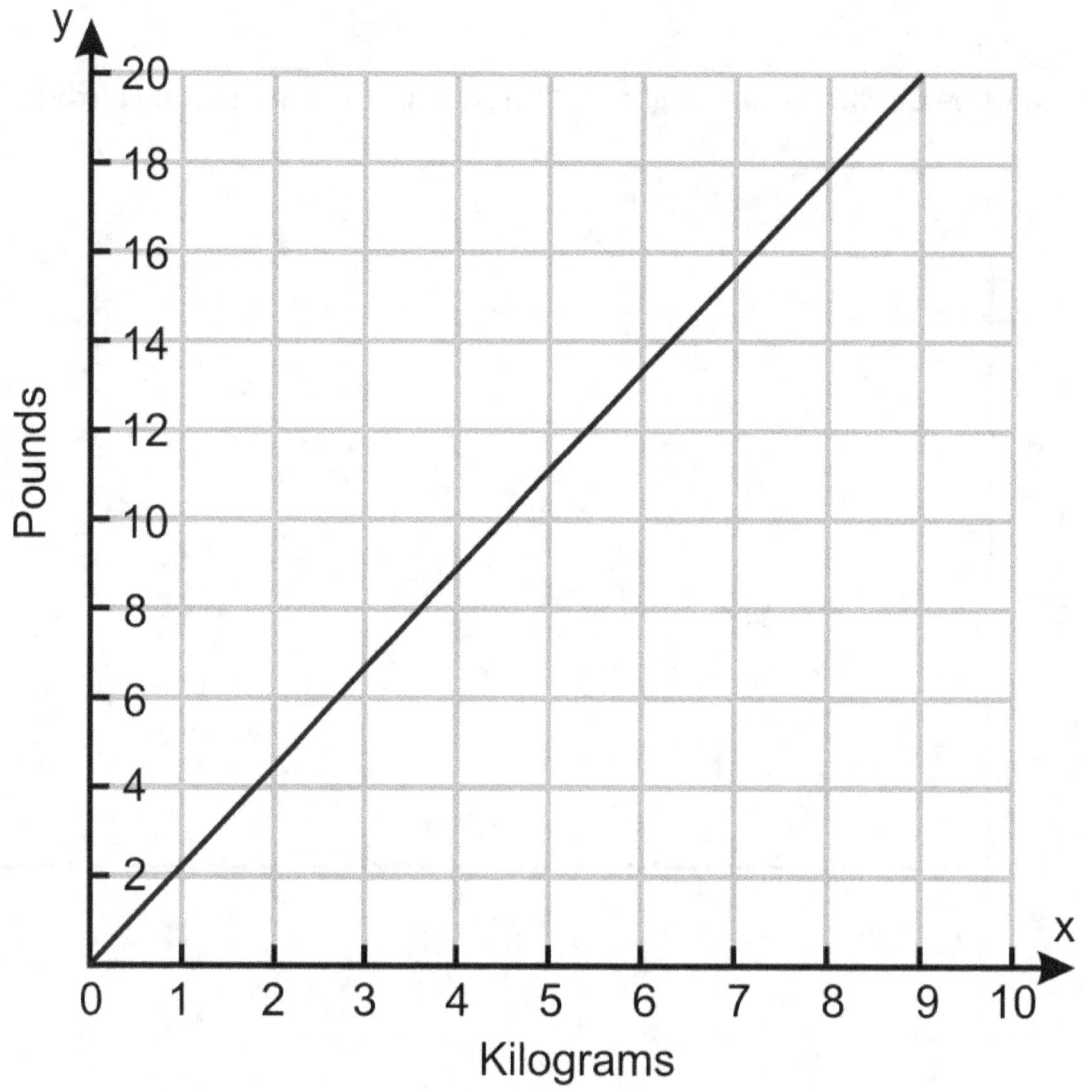

6. How many kilograms in 10lb? ____

7. How many kilograms in 20lb? ____

8. How many kilograms in 400lb? ____

9. How many pounds in 5kg? ____

10. Approximately, how many pounds would be in 10kg? ____

29.2 Depicting Data Pictorially

Pictograms are like bar charts, except they use images to show how many. Pictograms are also called pictographs. It is always important to look at the key to see how many each image is worth. Sometimes half an image is worth exactly half the full image, while sometimes it means less than the full image.

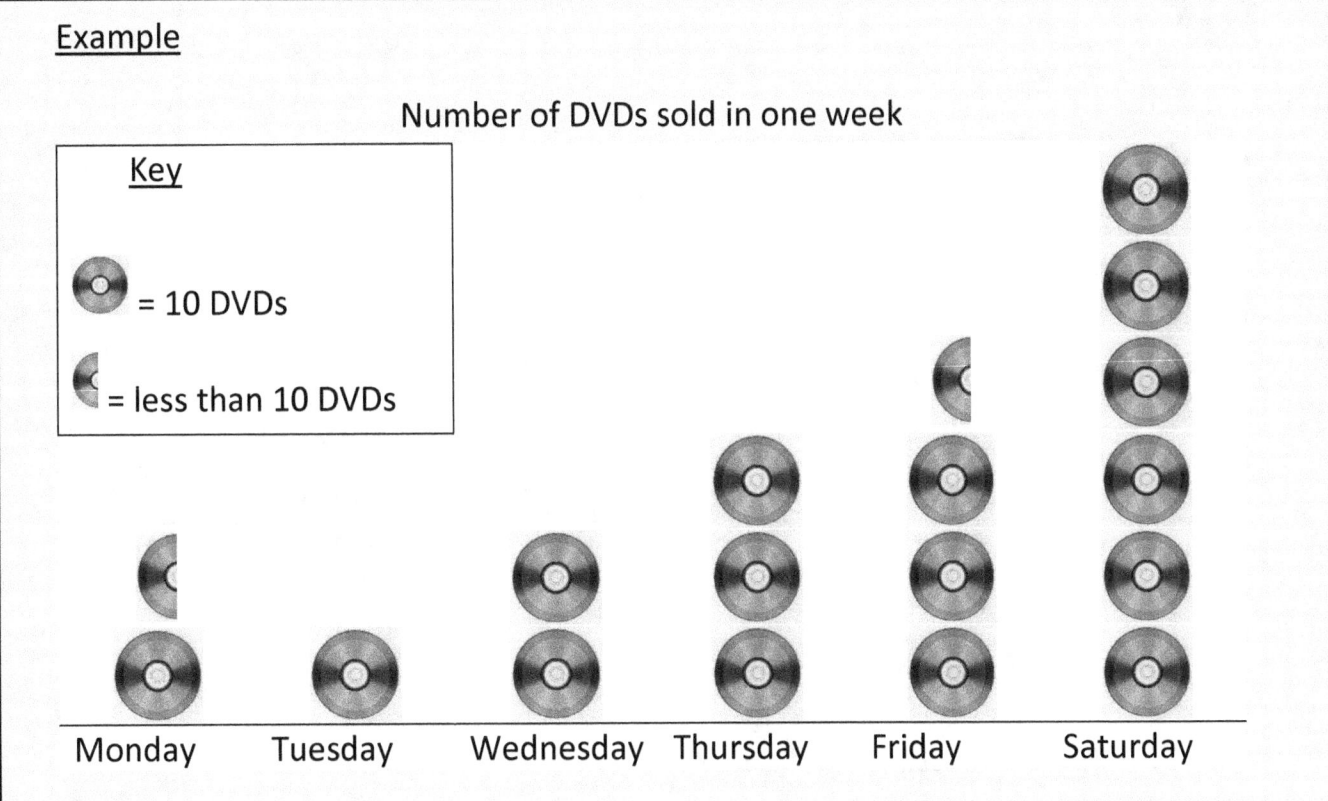

How many DVDs were sold on Wednesday? There are two pictures, each worth 10, so 20 DVDs were sold.

On what day were 32 DVDs sold? Thursday has exactly 3 pictures or 30 DVDs sold. Friday has 3 and a half pictures. Each half picture can be 1 to 9. Therefore the answer is Friday.

Exercise 29.2

Use the pictogram below to answer the following questions.

Key

 = 10 soft toys

 = 1-9 soft toys

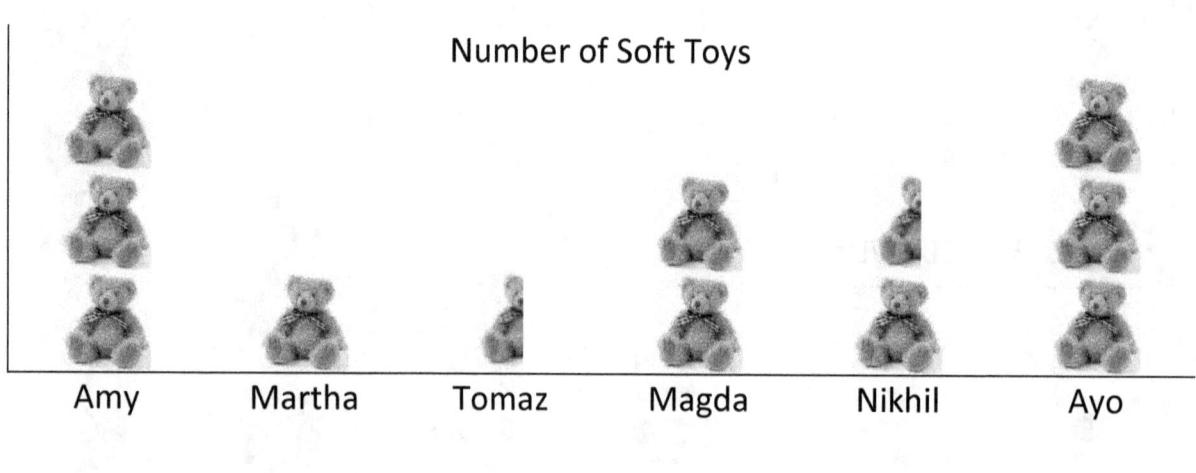

1. Who has the least number of soft toys?   _____

2. Who has exactly 10 soft-toys?   _____

3. If Martha gives half of her toys to Ayo, how many soft-toys will Ayo have?   _____

4. Who has 11 soft-toys?   _____

5. Which two children have the same number of soft-toys?

_____ and _____

Use the pictogram below to answer the following questions.

Key

✉ = 50 pieces of mail

◁ = 25 pieces of mail

Number of pieces of mail sorted by company mailroom

Monday	Tuesday	Wednesday	Thursday	Friday	Saturday
6 envelopes	1 half + 1 full = 1.5	4 envelopes (top is half)	3 envelopes	5 envelopes	2 envelopes

6. How many pieces of mail arrived on Monday? _____

7. How many pieces of mail arrived on Wednesday? _____

8. Which was the busiest day sorting mail? _____

9. How much more mail arrived on Monday than Tuesday? _____

10. How many pieces of mail were sorted in the whole week? _____

29.3 Venn Diagrams

Another way to illustrate data is in Venn Diagrams.

A set is a collection of things.

Venn diagrams put everything that belongs into a set into a circle. Circles overlap and the things that belong in both sets are in the overlap of the circles.

Anything that does not go in any circle is placed outside of the circles.

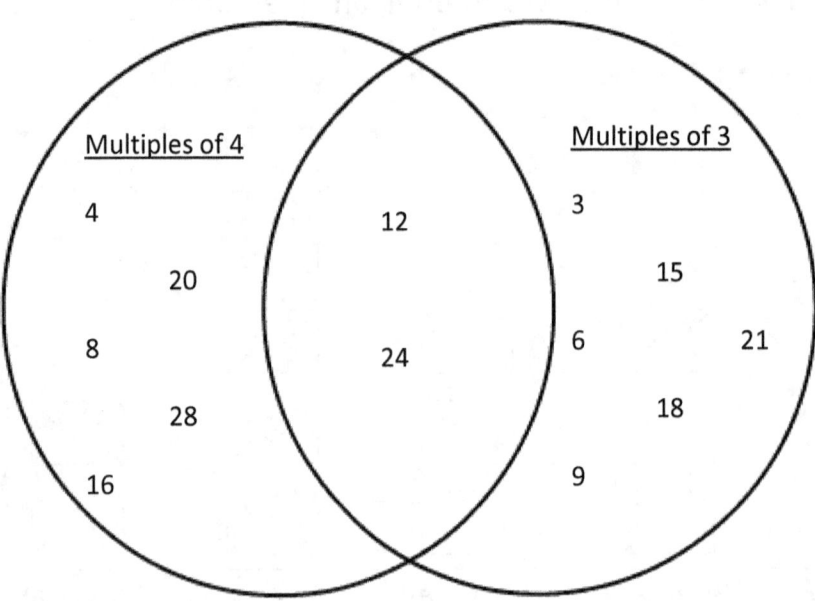

Exercise 29.3

1. Put the numbers 1 to 10 in the Venn Diagram below.

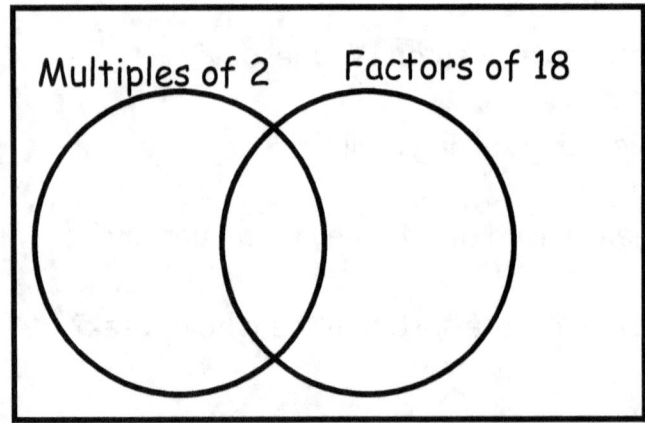

Use the Venn Diagram below:

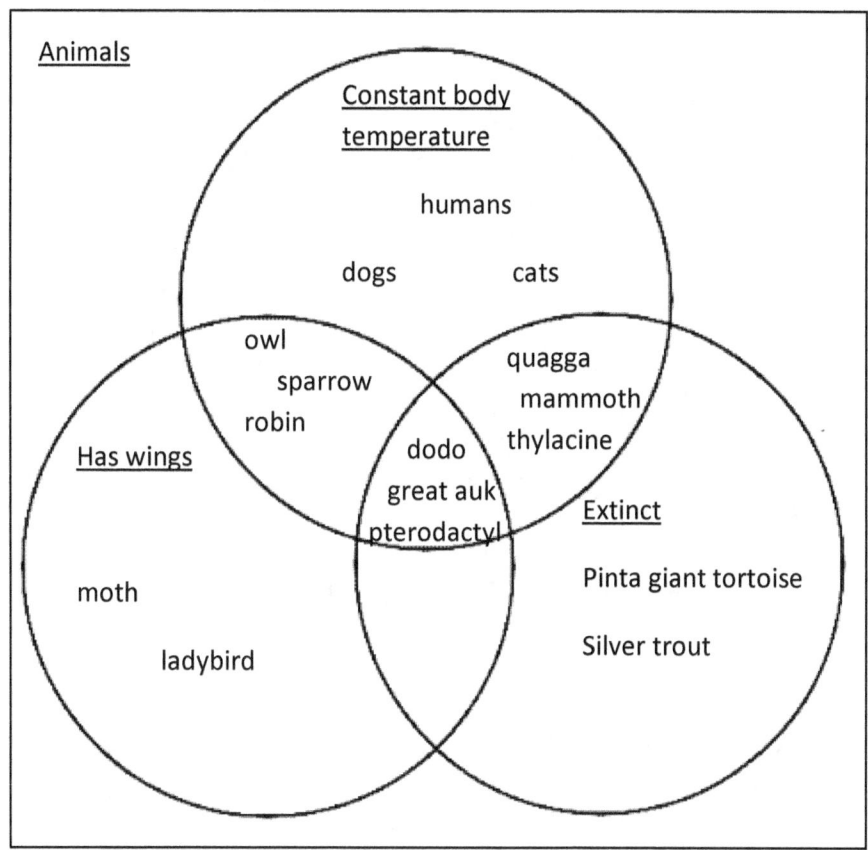

2. Name an animal in the Venn diagram that has constant body temperature and wings. _____

3. What do you know about Quaggas from the Venn Diagram?

_____

_____

_____

4. Name an animal that did not have a constant body temperature and is now extinct? _____

5. A pterodactyl is not normally considered a bird. Besides the dodo, what bird is extinct? _____

On a school mufti day the colours of the tops of 700 students is shown below.

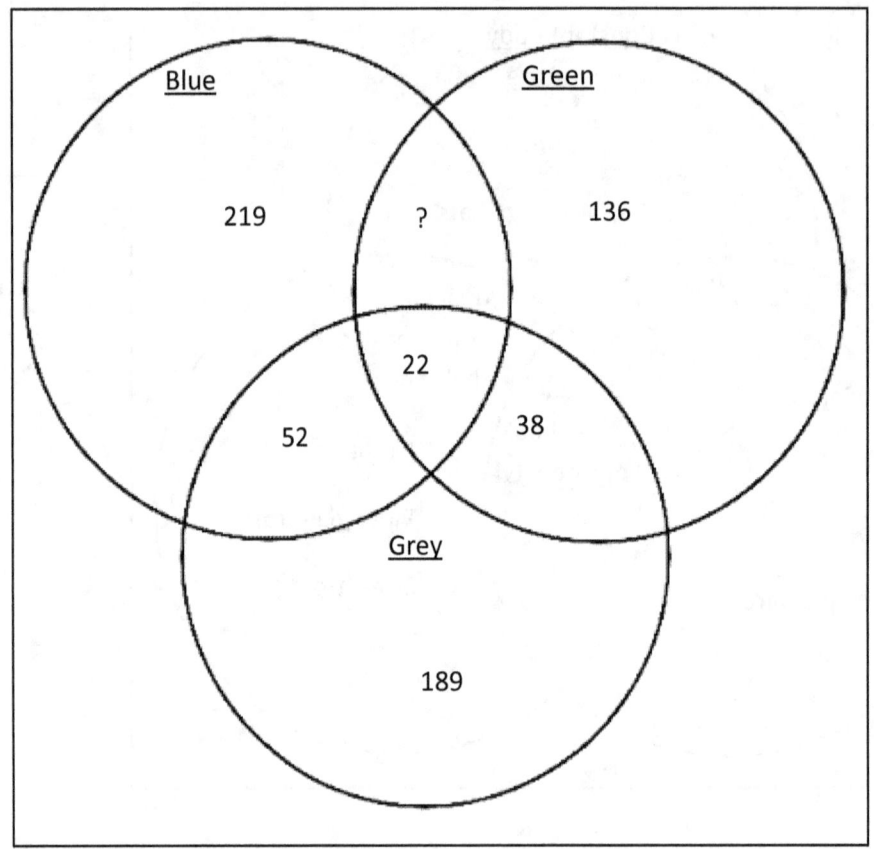

6. Blue and grey stripes were very popular. How many students wore blue and grey tops (with no green)? _____

7. How many students' tops contained both green and grey? _____

8. How many students managed to wear a top containing all three colours? _____

9. What percentage of students had some grey on their top? _____

10. What number does the "?" represent? _____

29.4 Flow Charts

Flow charts are another way of organising data or processes. You must start from the top and go in the direction of the arrows.

<u>Example</u>

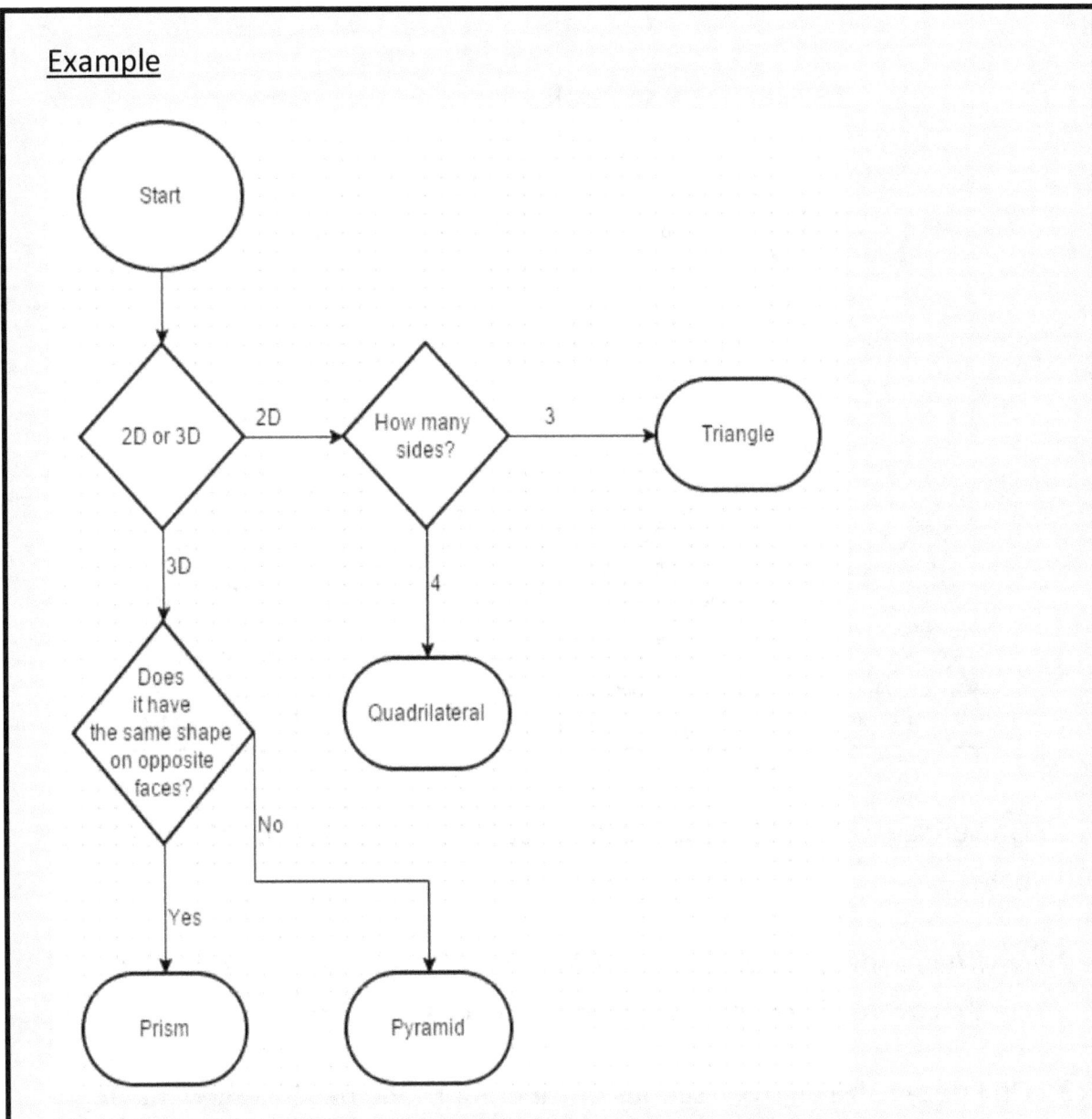

What is a shape called that is 3D and does not have opposite sides with the same shape? Following arrows from top, the answer is a pyramid.

Exercise 29.4

The following flowchart can be used to calculate the cost of an ice-cream.

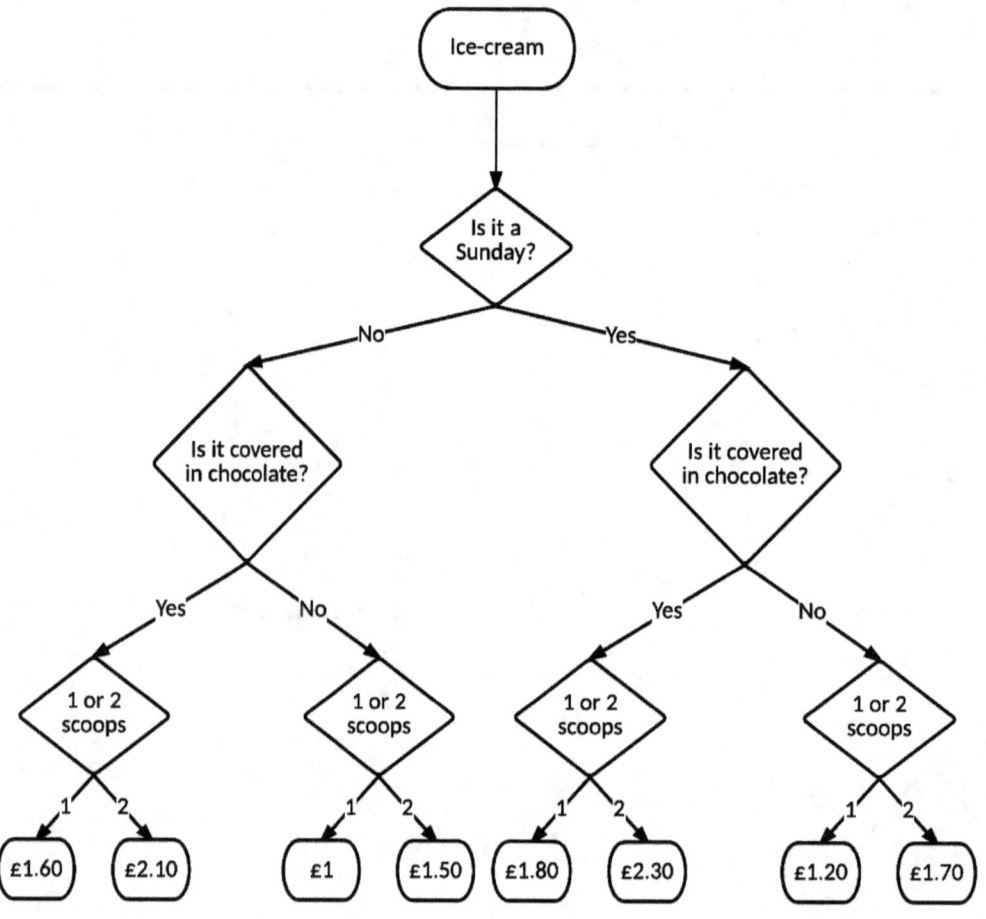

1. On Sunday, Uma buys 2 plain scoops of ice-cream. How much does Uma pay? _____

2. On Wednesday, Aditya goes to the park and buys a single scoop of ice-cream covered in chocolate. How much does it cost? _____

3. On Friday Faiza enjoys 2 scoops of ice-cream. She has $2. How much change does she get? _____

4. On Monday, Ria buys a single plain scoop of ice-cream. How much more would it have cost if she bought it on Sunday? _____

5. Asmita wants a chocolate covered ice-cream. How much will it cost on a weekday if she decides to get two scoops? _____

James is sorting out stationery.

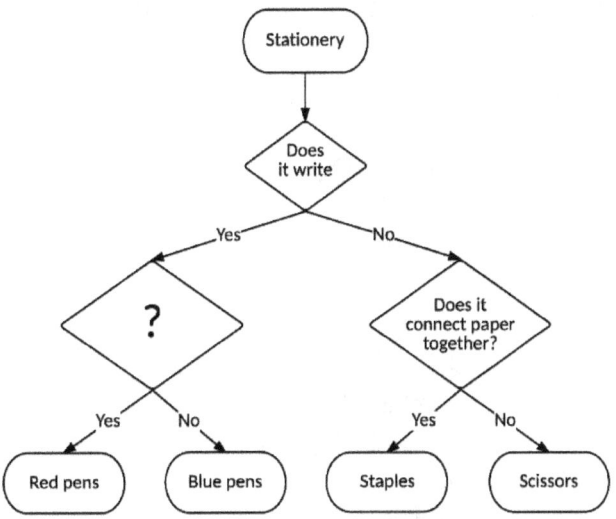

6. What should go in the empty box? _____
7. This flow chart only works for red and blue pens, scissors and staples. If James has a glue and uses the chart, what will it incorrectly be grouped with? _____

Adhil is sorting his clothes. He uses the flowchart below.

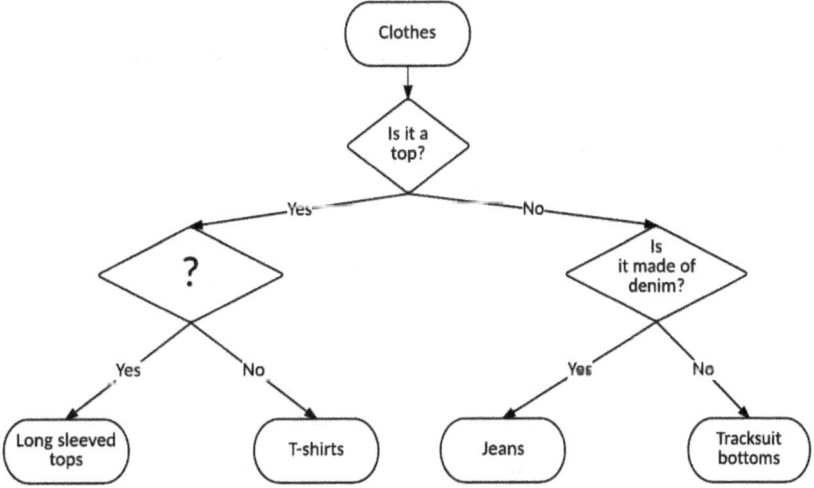

8. What should go in the empty box? _____

Simran is sorting some animals that are called arthropods, using the diagram below.

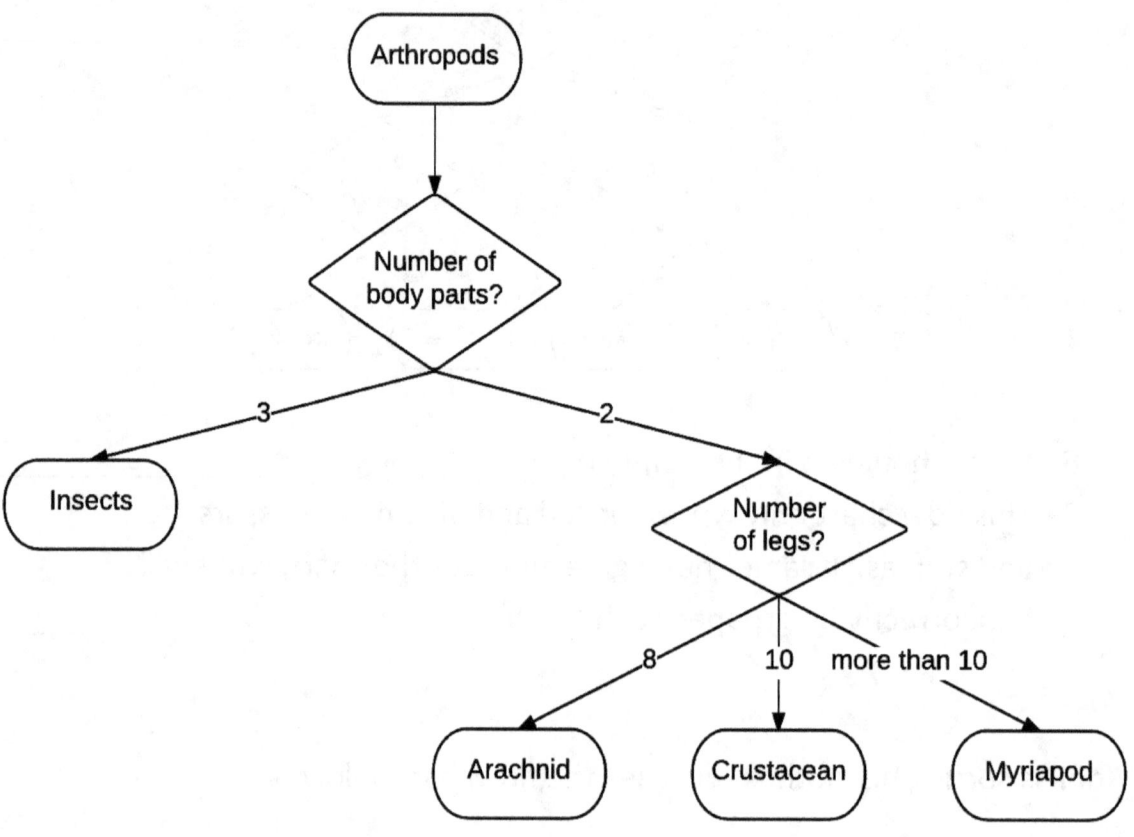

9. A tick has two body parts and four pairs of legs. What is it? _____

10. A woodlouse has 2 body parts and 10 legs. What is it? _____

Chapter 30: Algebra

30.1 Representing Unknowns with Letters.

If a number is missing, you can use the opposite maths operation to work out the missing number.

For example:

$2 + \square = 5$

We can work this out by doing subtraction $5 - 2 = 3$

Therefore, the missing number is 3.

Instead of using squares for missing numbers, letters can be used.

Example

The above question becomes

$2 + x = 5$

It is the same question. So, $x = 3$.

If we have a number and a letter together with no maths symbol, then it means times. So 2a means 2 times a. When we have a number and a letter together we always write the number first.

Example

$4x = 36$

$x = 9$ (as $4 \times 9 = 36$)

Exercise 30.1

Find a in the questions below.

1. 5 + a = 24 _____

2. 3a = 12 _____

3. 45 − a = 32 _____

4. 49 ÷ a = 7 _____

5. $\frac{a}{5} = 6$ _____

Find x in the questions below.

6. 7 + x = 15 _____

7. 12x = 132 _____

8. x − 23 = 40 _____

9. $x^2 = 16$ _____

10. $\frac{27}{x} = 9$ _____

30.2 Writing Math Problems using Algebra

In algebra we can use letters to represent different things. The letters are normally lower case but can be upper case too.

Example

1. If Mr Wiltshire bought two pencils (p) and an eraser (e) for $3, we could write this as: 2p + e = 3
2. In a fishbowl (f) there are 3 goldfish (g) and 2 zebra fish (z)
   f = 3g + 2z
3. A bus has 4 seats in each row (R) + one extra seat at the back where there is no aisle. Write an equation for the number of seats (S)
   S = 4R + 1

Exercise 30.2

Write an equation for the following.

1. The area (A) of a rectangle is the length (l) times the width (w)

2. The number of children (c) is the number of boys (b) and the number of girls (g)

3. The cost (c) of catching a taxi is a £3.50 call out fee and £2 per mile (m). Write the formula for calculating the cost of a taxi ride?

4. Petrol costs $1.30/litre. Write an equation for the cost (c) for v litres.

5. A chicken takes 40 minutes plus 30 minutes per kilogram to cook. Write a formula for the time taken to cook (c) in terms of the mass (m) or the chicken.

The following questions use these two shapes:

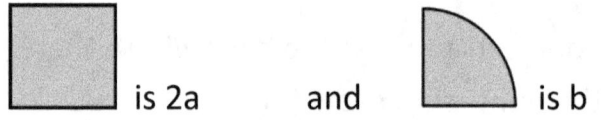 is 2a   and   is b

So:

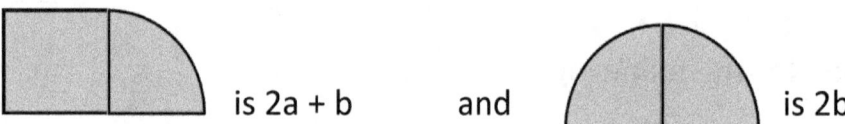 is 2a + b   and   is 2b

Write the formulae for the shaded shapes under the shape.

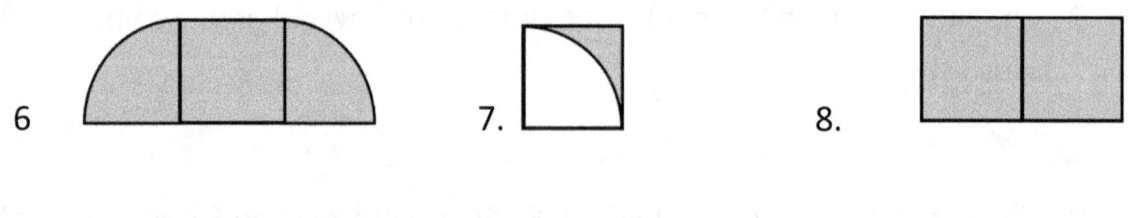

6.   7.   8.

9.   10.

30.3 Substitution

To substitute numbers into an equation, replace the letters with the numbers, the math symbols stay the same. Then do the question. Remember two letters together or a number and a letter together are multiplied.

Example

If a=2, b=3, c=5, d=7 and e=11, work out the following sum and give your answer as a letter.

$$ab + c =$$

This is the same as   a x b + c =

Replace the letters   2 x 3 + 5 =

Do the sum   2 x 3 + 5 = 11

Convert 11 to a letter   Answer is e.

Exercise 30.3

Give the answer to the following, *as a number*.

If a = 4, b = 3, c = 11, d = 8, e = 12

1. c + b =

2. e − b + a =

3. e ÷ b + c =

4. $\frac{b}{e} \times d =$

5. $b^a =$

Give the answer to the following, *as a letter*.

If a = 2, b = 6, c = 15, d = 3, e = 30

6. $\dfrac{b}{d}$ = _____

7. cd – e = _____

8. b^a – e = _____

If m = 23, n = 16, p = 5, q = 50, r = 13, s = 3

9. (q + p) ÷ (n – p) = _____

10. $\dfrac{m+n}{s}$ = _____

30.4 Using Formulae

Here are some formulae you will come across:

Area of a rectangle is length times width
A = l x w (or A = lw)

The perimeter of a rectangle is
P = 2l + 2w

The volume of a cuboid is
V = l x w x h (or V=lwh)

Example

What is the volume of the cuboid shown?

V = l x w x h

= 8 x 6 x 5

= 240 cm^2

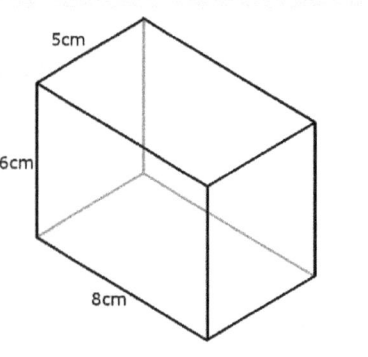

Exercise 30.4

Now try these, using the diagrams below:

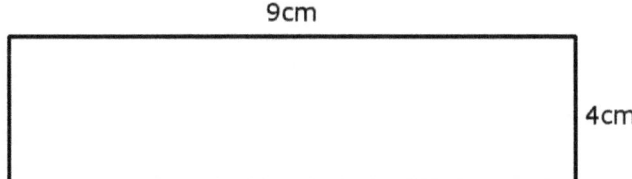

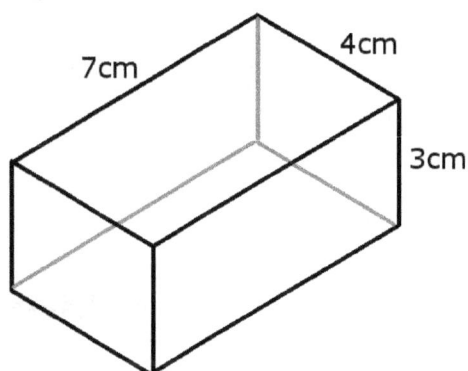

1. What is the perimeter of the rectangle?

2. What is the area of the rectangle?

3. What is the volume of the cuboid?

4. A taxi charges according to the formula:
$$c = 3.5 + 0.5t$$
where c is the cost in pounds and t is the time of the journey in minutes. Calculate how much it will cost for a 10 minute journey.

5. The formula for the cost (c) of a newspaper advert, including the $45 set up fee, is: (height of ad is given as i)
$$c = 45 + 5i$$
Calculate the cost for an ad of 25 inches.

6. How long it takes a roast to cook can be calculated by the formula:
$$t = 40 + 30m$$
Where t is the time in minutes, and m is the mass of the roast. How long does it take to cook a 6kg roast?

7. The cost of items in a sale can be calculated by the formula:
$$s = \frac{3n}{4}$$
where s is the cost in the sale and n is the normal cost. What is the cost of an item that was $20 before the sale?

8. The cost (C) of hiring a hall is determined by the formula:
$$C = 250 + 10G$$
Calculate the cost of hiring the hall on Saturday for 200 guests. (G is the number of guests)

9. The formula for speed is:
$$s = \frac{d}{t}$$
Where s is the speed, d is the distance and t is the time.
Calculate the average speed travelled, when Derek travels 342km in 6 hours. Can you work out the correct units?

10. The formula for energy due to the height of an object is shown below.
$$E=mgh$$
Where E is the energy in Joules, m is the mass in kg, g is the gravitational strength and h is the height in metres. How much energy does a 200kg rock have at a height of 10.75m? (gravity =10)

30.5 Algebraic terms

Coefficients and Variables

In math, the letters are called variables and a number in front of the variable is called a coefficient. A number without a letter is called a constant.

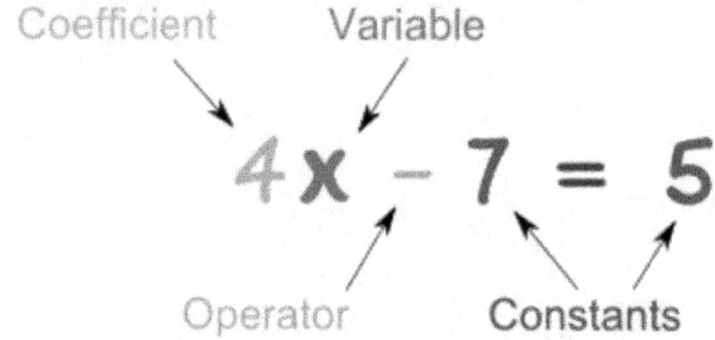

Like and Unlike

Variables that have the same letter and the same index, are said to be like. Variables that have different letters are said to be unlike. If a variable has the same letter but a different index, then it is unlike.

For example: 4d and 8d are like

        5f and 6g are unlike

        3a and 4ab are unlike

        9y and $3y^2$ are unlike

Exercise 30.5a

Write whether the following is a coefficient, a variable or a constant

1. The x in 3x   _____
2. The 7 in 7a   _____
3. The 4 in 4x + 2   _____
4. The b in 8b + 2   _____
5. The 8 in 2x − 8   _____

Write L if the terms are like and U if the terms are unlike.

6. 8t and 31t _____

7. 7w and 11w^3 _____

8. 8r and 11r _____

9. v and 8j _____

10. 6p^2 and 9p^4 _____

We can combine like terms.

For example:

3a + 4a = 7a

8ab – 5ab = 3ab

> If a stands for apples:
>
>
>
> then 3 apples + 2 apples = 5 apples
>
> or 3a + 2a = 5a

So:

3a + 8b +4a – 5b = 7a - 3b

3e^2 -2e + 6e –e^2 + 3 = 2e^2 +4e +3

7h – 3j = 7h – 3j

7a + 2ab + 6ab + b = 7a + 8ab + b

Exercise 30.5b

Simplify the following expressions:

1. 5a + 3a = _____

2. 9w – 2w = _____

3. 12a + 3b – 7a -2b = _____

4. 8d^2 + 3d + 2d = _____

5. $12x^3 + 3x^3 + 2x^2 + 4x + 11 =$ _____

6. $15z^2 - 9z^2 + 4z - 3z + 4 =$ _____

7. $45b^2 - 4b + 7t - 5t + s =$ _____

8. $21w + 9wx + 6wx - 3x =$ _____

9. $78q - 6qr + 11qr - 8q =$ _____

10. $8s^3 + 4s - 3s^3 - 4s =$ _____

Chapter 31: Using Algebra

31.1 Multiplying Terms

A number or term multiplied by itself is that number squared.

For example:

5 x 5 = 5^2

a x a = a^2

Numbers multiply as normal.

For example:

3a x 2a = $6a^2$

If two numbers or a number or letter are together with no math symbol, then they are multiplied. So 4a means 4 lots of a or 4 multiplied by a.

For example:

a x b = ab

4w x 3y = 12wy

6s x 3st = $18s^2t$

Exercise 31.1

1. s x t = _____
2. 2a x b = _____
3. 4w x w = _____
4. 7y x 2y = _____

5. 9z x 3yz = _____

6. 6s x 3s x t = _____

7. y x 5yz = _____

8. 11 x 2a = _____

9. 0.5r x 8rs = _____

10. 9t x 3tv = _____

31.2 Dividing Terms

Numbers and letters can be divided individually. Two of the same letter cancels each other out. A squared letter divided by the letter, is just the letter.

For example:

$a^2 \div a = a$

$6y^2z \div 2yz = 3y$

Exercise 31.2a

1. $2n^2 \div 2 =$ _____

2. $12rs^2 \div 3r =$ _____

3. $12rs^2 \div 4s =$ _____

4. $\dfrac{14p^2}{2p} =$ _____

5. $\dfrac{11a^2b}{ab} =$ _____

6. $5n \div n =$ _____

7. $\dfrac{18x^2y}{3x} =$ _____

8. $\dfrac{21b^7}{3b^3} =$ _____

9. $\dfrac{12s^2t}{4stu} =$ _____

10. $\dfrac{15x^5y^2}{5} =$ _____

Combining multiplication and addition

It may be necessary to both add and multiply when simplifying an expression. When doing so, remember the order of operations.

For example:
3m + 2m x 5m =
We do multiplication before addition, so we do 2m x 5m first
2m x 5m = 10m²
We are now left with
3m + 10m²
This cannot be simplified further, so the answer is
10m² + 3m

Exercise 31.2b

1. $4 \times (3s + 2s) = $ _____

2. $6a \times 2a + 3a^2 = $ _____

3. $(5e + 3e) \times 2e = $ _____

4. $6w \times w + w^2 = $ _____

5. $3d \times (4d + 2d) = $ _____

6. $3a \times 2a + 2ab = $ _____

7. $3a \times 5b + ab = $ _____

8. $5s \times 2st + 7t = $ _____

9. $6q \times 7q - 2q = $ _____

10. $9y(12y - 4y) = $ _____

> Question 1 could also be written as $4(3s + 2s)$.
>
> Similarly, question 3 could also be written as $2e(5e + 3e)$

31.3 Balancing an Equation

To balance an equation, all you need to remember is that whatever you do to one side of the equation to the other side you do **exactly the same.**

To solve an equation:

1. Get all the terms with the letter on the same side.
2. Get the term with the letter on its own.
3. Get the letter on its own.

There are two methods to do this:

- Balancing method
- Change side, change sign.

Example

Solve: $5x + 3 = 48$

Method 1

$5x + 3 = 48$	To get 5x on its own need to start by getting rid of the +3.
$5x + 3 - 3 = 48 - 3$	To get rid of the +3, need to -3. Whatever I do to one side, to the other side I do **exactly the same**.
$5x = 45$	5x is the same as 5 times x, so to get just x, I need to divide by 5. Whatever I do to one side, to the other side I do **exactly the same**.
$\frac{5x}{5} = \frac{45}{5}$	
$x = 9$	

Method 2

$5x + 3 = 48$	To get 5x on its own, start by getting rid of the +3.
$5x + 3 = 48 - 3$	So, move the +3, but when it is moved to the other side, it becomes -3.
$5x = 45$	Now, move the 5 to the other side, but when it moves it becomes the opposite operation so it becomes divide.
$x = \frac{45}{5}$	
$x = 9$	

Exercise 31.3

1. x + 11 = 22 x = _____

2. 11x = 33 x = _____

3. 5x + 3 = 38 x = _____

4. 9x − 14 = 22 x = _____

5. $\frac{x+3}{2} = 9$ x = _____

6. $\frac{n}{4} = 8$ n = _____

7. $\frac{5n}{7} = 10$ n = _____

8. 4z + 7 = 13 z = _____

9. 15z − 17 = 43 z = _____

10. 0.5z + 3 = 5 z = _____

31.4 Creating Formulae

There are two main reasons that creating formulae could be useful in the 11 Plus exam.

- Working out the nth term.
- Solving word problems.

<u>Working out the nth term</u>

Given a series of numbers the difference between each number is what you multiply by and then you add or subtract a number to get the first term.

Example

The shapes below are made from matchsticks.

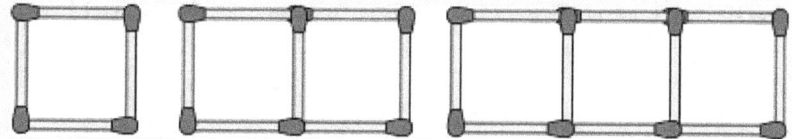

How many matchsticks would I need for the 20th shape?

Start by counting the number of matchsticks in each shape.

Term number	1	2	3
	4	7	10

Each number increases by three, so if we use t for term, the first part of the formulae will be:

No of matchsticks = 3t

The first term is 4.

Using the first part of the formula above 1 x 3 = 3. I need to add 1 to make it 4. So my complete formula is:

No of matchsticks = 3t + 1

To find the 20th term, substitute the number 20 for t.

No of matchsticks = 3t + 1

$= 3 \times 20 + 1$

$= 61$

Problem solving

Using algebra can be the easiest way to solve some word problems.

Replace items you want to work out with a letter (normally a related letter) and then solve.

Example

Tori has 36 apples and oranges in a fruit bowl. She has three times as many apples as oranges. How many oranges does she have?

If we refer to the number of oranges as F, then the number of apples is 3F.

No of apples + no of oranges = no of fruit

3F + F = 36

4F = 36

F = 9

She has 9 oranges.

Example

If a number is multiplied by six and 5 is added the answer is 47, what was the number?

Replace the number with x

6x + 5 = 47

Solve as before

6x + 5 = 47

6x + 5 − 5 = 47 − 5

6x = 42

$$\frac{6x}{6} = \frac{42}{6}$$

x = 7

Example

Akshayan has a younger sister Brianna. Akshayan is four years older than Brianna. The sum of their ages is 12, how old is Brianna?

Let B = Brianna's age

Akshayan's age is B+4

The sum of their ages is 12.

So: B + (B+4) = 12

2B + 4 = 12

2B + 4 − 4 = 12 − 4

2B = 8

B = 4

So, Brianna's age is 4

Akshayan's age is B+4 = 8

Example

Friday after school Shadrak starts reading a new book. On Saturday, Shadrak reads 25 more pages than he read on Friday. He reads 105 pages altogether. How many pages did he read on Saturday?

Let the number of pages he read on Friday be x.

So the number of pages read on Saturday = x+25

x + x + 25 = 105

2x + 25 = 105

2x = 105 − 25

2x = 80

x = 40

Pages read on Saturday = x + 25

= 40 + 25

On Saturday he read 65 pages.

Exercise 31.4

Write the equation for the number series below and work out the required term. Use the letter x when writing the equations.

1.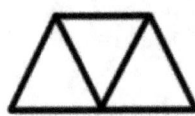

   equation: _____, 10th term: _____

2. 5, 11, 17, 23, ...
   equation: _____, 12th term: _____

3. 25, 27, 29, 31, ...
   equation: _____, 15th term: _____

4. -4, -1, 2, 5, ...
   equation: _____, 100th term: _____

5. -20, -15, -10, -5, ...
   equation: _____, 20th term: _____

Use algebra to work out the following problems. Write the equation used and your answer.

6. A number is tripled and the 56 is added. If the answer is 62, what was the number?
   equation: _____, number: _____

7. A number is multiplied by 5 and then 22 is subtracted. If the answer is 78, what was the number?
   equation: _____, number: _____

8. When 17 is added to a quarter of this number the answer is 47. What is the number?
   equation: _____, number: _____

9. In two years time Rhiannon will be half Dominic's current age. If Dominic is 12 how old is Rhiannon?
   equation: _____, number: _____

10. Jane spent $22 more on Jeans than she did on a pair of shoes. She spent $60 altogether. How much did she spend on the shoes?
    equation: _____, number: _____

Puzzle Page 5

Why did the Romans not find algebra very challenging?

$5782 + 364 + 97 - 12 =$ 6231   c

Cats and dogs are in an animal shelter is a ratio of 5:4. There are 70 cats.

1. How many dogs are there? _____ ___
2. How many animals are there altogether? _____ ___

Taryn is 6 times older than Myra who is 2 times older than Kate. Their total age is 90.

3. How old is Taryn? _____ ___
4. How old is Myra? _____ ___
5. How old is Kate? _____ ___

A restaurant has a special for one main, one side and one desert for $10. There are 12 mains, 4 sides and 5 deserts in the offer.

6. How many possible combinations are there? _____ ___
7. If three of the deserts contain nuts. How many nut-free options are there? _____ ___
8. A bag contains blue, green and red balls. The chance of pulling out a red ball is $\frac{1}{3}$. If there are 215 green balls and 245 blue. How many red balls are there? _____ ___
9. Solve for x. $5x + 3 = 303$ _____ ___
10. Solve for x. $\frac{x}{3} + 2 = 5$ _____ ___
11. Solve for x. $6x = 180 + 36$ _____ ___
12. Solve for x. $2x + 3x - 5 = 10$ _____ ___

Work out what letter each of your answers represents below:

6	240	36	56	12	3	72
w	r	o	s	e	x	t

6231	126	9	96	230	60
c	y	l	a	h	n

Now write the letter above the question numbers, below to answer the puzzle.

__   __ __ __   __ __ __ __ __ __
12   5 7 1   7 10 5 7 2 1

__ __ __ __ __   __ __ __
5 11 6 3 8   3 4 9

www.ingramcontent.com/pod-product-compliance
Lightning Source LLC
Chambersburg PA
CBHW050716090526
44588CB00015B/2339